THE COMING TECHNOLOGY TSUNAMI

A Personal History of the Future
BY DR. ROCCO LEONARD MARTINO

PUBLISHED BY

BlueNose
PRESS, INC.

Printed in the United States of America
Published January, 2017

Author Portrait: Bachrach
Full Cover & Title Page: Hedy Sirico
Interior Layout Design: Joseph A. Martino

For more information on this title please visit:
www.BlueNosePress.com

ACKNOWLEDGEMENTS

A book of this nature cannot be written without the support of many people. As I wrote, many friends from the past, no longer with us, whispered in my ear "Remember when..." I remember arguing circuit design while walking through snow banks with John Mauchly; disagreeing with Grace Hopper on how best to represent data structures in a coding environment; reminiscing with Kay Mauchly on programming ENIAC; comparing notes with Pres Eckert on hi-fi music, and hi-fi circuits in computing; arguing ethics with Jacques Maritain; probing the limit of the universe with Leopold Infeld; envisaging spaceflight with Gordon Patterson; building a University with Doug Wright; and molding the minds of young men at the University of Waterloo. But best of all, I remember the long discussions with my Jesuit buddies in Rome and many cities of the world. Del Skillingstad, Jake Laboon, John Snyder, Patrick Heelen, John Blewett, Joe Pittau, and Mike Smith will be with me always, together with my living Jesuit buddies – Domenic Maruca, Nick Rashford, Arturo Lozano, George Aschenbrenner, George Bur, John Horn, Steve Pisano, and Gerald O'Collins.

I strongly recommend the web as a source of the latest progress in Technology. In particular, I recommend the writings of Peter Diamandis. His weekly reports of technology progress under the title 'Abundance Insider' is must reading.

I am particularly indebted to Dr. Jack Schrems, Professor Emeritus of Political Science at Villanova University here in Pennsylvania, for his counsel, insights, and editorial comment in my attempt to bring this whole subject matter to the attention of the reader. More directly, Anne Condello and Christine Monigle were instrumental in transcribing my dictations, editing my structures, and commenting on my ideas. Hedy Sirico designed the cover and the title page. My second son, Joseph Martino was instrumental in formatting and getting this book published.

My deepest thanks to all of you.

DEDICATION

I dedicate this book to all those who made it happen. As I sit here thinking back over my own life, all 87 years to date, I remember so many people and events that are the cornerstone of the developments I write about in this book. Believe me, it was an adventure. I remember the pioneers, the entrepreneurs, the hotshots, the starry-eyed, the scoundrels, the opportunists, and the many good people who supported change, used it, and made it a prominent part of their lives, and that of everyone else. But we cannot live in the past. We must use the past to create the future. Hence, I dedicate this book to those who will make the future happen. The world will benefit greatly from those with "fire in the belly", provided greed and political power are not allowed to curtail innovation and discovery.

Symbolically, I dedicate this book to my grandchildren as I see them taking their role in making the dreams of old men like myself come true.

REVIEWS

Listed below are condensed versions of the reviews for this book. The full reviews can be read on the BlueNose Press website at the following URL: http://www.bluenosepress.com/reviews.html

#

Dr. Martino's predictions are utterly fascinating but also somewhat frightening, given that almost of all of them are imminent. His comments on the connected nature of the world and what it means for technology is entirely correct. His arguments for the emerging ubiquity of solar power, electric vehicles, driverless cars, and robotics all seem to me to be right on the money. Many would agree with Dr. Martino's 10-year prediction.

The chapter on space and the computer made for enjoyable reading given my professional interests. The historical description that Dr. Martino presented provided an interesting way of establishing the "time constant" for our accomplishments in space. We are no doubt going to get to Mars in the relatively near future.

Christopher J. Damaren, PhD
Professor and Director, University of Toronto
Institute for Aerospace Studies

Dr. Martino does it again with a provocative message, yet one that will awaken the true educator's love for personalized learning and innovation! Nothing could be more timely than this "Tsunami alert" for higher education (and perhaps all of education). At a moment when higher education finds itself at a cross roads - seeking solutions that address the value proposition, creating efficiencies to control costs, transforming faculty ways of doing, and providing an education that prepares for life while pressured to ensure marketable job skills - Martino's message, is our Tsunami alert to prepare, to create opportunities for the academy to gather and explore thoughts and possibilities, predictions and visions for this future technological transformation. Just as a Tsunami advisory motivates and directs "energy and life-saving actions," Dr. Martino's book provides advice, motivates proactive thinking, and prods us to look beyond the immediate moment, embrace the inevitable changes and develop innovative solutions that address the future.

Ultimately, to ignore the possibilities brought forth by scientific and technological advances is to ignore reality; to move forward drawing on those possibilities to personalize learning, to engage and transform the learner through the lens of ethics and morals is to bring Education to its most sacred form and place. Dr. Martino's book provides advanced warning that technology cannot be ignored. Who would want to if it is helpful in bringing education to its rightful and esteemed place in creating a better world for all? A must read for all involved in formal education today.

Rosalie M. Mirenda, PhD
President, Neumann University

Colonization trip to Mars in the next ten years? Hard to imagine. Yet remember, the self-driving car was only imagined by engineers, and not the rest of us, just a few years ago. So now an engineer-rocket scientist who assisted in the launching of space science sixty years ago, tells us of a tsunami of technology in the next ten years. Robots in that time? Already my local trash weekly pick-up has gone from a five-person crew to one person. Jobs will be lost. Yes! But look at the bursting size of universities and the computer industries for confirmation that new opportunities are already with us. This new book by Dr. Rocco Martino gives us a glance of what the future will be like.

John J. Schrems, PhD
Professor Emeritus of Political Science
Villanova University

Dr. Rocco Leonard Martino presents a challenge to the aphorism that the future is an unknown; the present is chaos and the past is the only certitude we can examine.

This book is not a work of science fiction, but rather an utterly believable and understandable insight into how our world will change over the next 10 years. Dr. Martino foresees the new world order in terms of our passage on this planet becoming better for mankind as technology brings forth new means to improve the quality of life we might hope to achieve.

In applying his forecasts to contemporary events, Dr. Martino predicts a shattering decline in the use of petroleum-derived products, seeing a future in which oil is a lubricant rather than a fuel, with the exception of jet fuel, thereby causing significant destabilization in those parts of the world that rely upon its export as their lifeblood.

For some who read this book the future will remain an unknown, but for those who wish to envision what it will be, this book is a fine place to begin that journey. Upon finishing reading it, one comes away without fear of what will transpire in the years to come, but rather with hope that the present limits of our abilities will pale as we embrace a new world, replete with technology, but also with recognition of our growing ability to profit in making life better for all mankind.

Henry Lane Hull, PhD
Retired Professor of History
University of Alabama in Huntsville

WORKS BY ROCCO LEONARD MARTINO

Fiction
The Cross of Victory
Christianity: A Criminal Investigation...
The Resurrection: A Criminal Investigation...
9-11-11: The Tenth Anniversary Attack
The Plot to Cancel Christmas

Nonfiction
Memories: Volume I - Stories for My Grandchildren
Memories: Volume II - Scientist and Writer
Memories: Volume III - Changing the World
Rocket Ships and God
Walking Around the Neighborhood
People, Machines, and Politics of the Cyber Age Creation
Finding the Critical Path
Applied Operational Planning
Allocating and Scheduling Resources
Critical Path Networks
Resources Management
Dynamic Costing
Project Management
Decision Patterns
Decision Tables with Staff of MDI
Information Management
Integrated Manufacturing Systems
Management Information Systems
MIS Methodology
Personnel Management Systems
IMPACT 70s with John Gentile

TABLE OF CONTENTS

PREFACE

It took me 70 years to write this book. I enjoyed every minute of it. I've already written some history of the past in other books. This book is my attempt to write the history of the future. I think it will all be possible. In fact, I think my book is an understatement of the unbelievable advances of the next 10 years as we reap 70 years of plowing the fields of science and engineering as we ultimately created the totally wired world.

The reader may wonder if some of my assertions are cut from whole cloth – figments of my imagination. I can assure you every projection is based on solid facts from multiple sources. I am scientist. I decided not to clutter the book with tedious references of proof. The proof of my projections will unfold in actuality shortly.

Who am I to make these predictions? A brief bio is at the back of the book. More importantly, I was there close to ground zero from the beginning. I wrote my first computer program in 1949. I used one of the world's largest computers for aerospace problems beginning in 1952. I discovered the heating characteristics of space vehicles returning to earth. I worked with the inventor of radar, Sir Robert Watson Watt; was in partnership with Dr. John Mauchly, the co-inventor of the first computer, the ENIAC; and worked with Admiral Grace Hopper on automatic programming techniques. In my own company, XRT, I invented the concepts of real-time dual storage of transactions, and the extensions to disaster tolerant systems to continue operations in the event of natural or man-made disasters; I created the first electronic funds transfer system, the first online real-time trading system, the first 100% uptime system, and the first Treasury Workstation. By the mid-1990's, systems I designed for clients were moving over three trillion dollars per day without any penetrations of any kind. Working with physicians, I developed remote diagnosis systems. I served many of the largest corporations and government entities in the world on many major projects. I invented the Smartphone, which I called the CyberFone, and filed for the patents on May 19, 1995. I filed for patents on the CyberWatch and medical diagnosis systems using nanobodies. I have been in the forefront of technological development for 65 years. That experience is what gives me the basis of my vision of the next ten years.

It will be disruptive. That is because it will change everything. The status quo will be turned upside down. But then again, this is not a buggy whip world. The horse and buggy may be romantic on Christmas cards or on the rare excursion into a ski resort. By the same token, the fireplace may be cozy, comfortable, and romantic, but certainly not as effective as a thermostatically controlled central heating system. I hope this book will be a challenge to you, the reader. No matter what your age, I hope you will pick up the gauntlet that is possible and move it forward to achieve the maximum potential. I see the next 10 years as the beginning of the Golden Age of Human Kind.

CHAPTER 1
The New Connected World:
The Beginning

A tsunami is created when an earthquake or some other cataclysmic activity occurs under the surface of the ocean. Once initiated, no matter how small it might appear at the beginning, it travels at enormous speed across vast distances, ever increasing in strength. On landfall, the wave heights are thirty feet or more and proceed to wipe out everything before them. Nothing is the same after such a tidal wave has passed through a built-up city or town.

Once initiated a tsunami is inevitable. It cannot be diverted, diminished, or stopped. The impact cannot be avoided.

A technology tsunami is about to strike the world. It is inevitable and cannot be avoided. It started on February 14, 1946 when ENIAC, the first digital computer, was demonstrated publicly at the University of Pennsylvania in Philadelphia. In the 70 years since, more and more devices and applications have added to the power of the tsunami. From this single computer, the number of computers today exceeds the population of the earth. Major new effects of the tsunami are about the strike the world. There will be many casualties, with major changes in the way we live and work. Nothing will be the same afterwards. The first casualty will probably be your job and your way of life. It could

3

destroy your career and life, or it could be the beginning of new opportunity and challenge. Get ready!

Within the next ten years, the following will most likely occur. The speed of implementation of each of these will vary, but they will most likely all be completed within the decade beginning today.

1. The Smartphone will be the ubiquitous element of life virtually mandatory from birth. The cost will be significantly reduced, and we will most likely wear our Smartphones as a part of our dress. The unit will become small enough to fit in an ear if you choose to so wear it.

2. The entire earth will be continually connected, with Wi-Fi, internet, telephony, messaging, GPS, voice and video available at all times, everywhere.

3. The cost of communication will be virtually zero.

4. Billons of new job seekers will enter the employment stream; and most of these will be unskilled. They will require extensive education and training.

5. The oil and gas industry will be diminished in importance. It will move to a significant reduction in power. Like coal, oil and gas will become a commodity for lubrication and not vital for providing power or transportation on earth. Jet fuel will most likely be the only fuel produced. The

disruptions in the economies of many nations will be enormous.

6. Power will be produced more and more by solar techniques. The movement will begin to have every home generating solar power, together with vast farms of solar power generators. The cost per kilowatt hour generated will be pennies. Major battery areas will be available in the electric grid for stand-by power in the absence of sunlight.

7. Cars will become increasingly electric. Most new production units will be driverless. Accidents will diminish dramatically. Hit and run crimes will be eliminated. Driverless taxis will dominate the transportation milieu; the car industry will be decimated, along with the jobs in the industry. New personal travel vehicles that are small in size, and are capable of operating on specially created highway networks will become standard items for the individual.

8. Millions of jobs will be eliminated - at least 25%, if not 50%, of the existing rote, mechanical, or low-skill jobs. These reductions will apply even to those most skilled such as doctors, lawyers, and engineers.

9. Check-out clerks in stores and malls will no longer be needed as sensors will perform the same functions. Shoppers will make their selections, and either carry them out or have them delivered; and just leave the store. The

sensors will establish the charges and billing with delivery instructions, if needed, and they will be generated automatically.

10. There will be at least one robot in each home in the United States and most of the developed nations of the world. Robots will also begin to be in use in homes and industries of the developing world. Robots will replace many of the jobs currently performed by humans. Robots will perform many jobs not being performed at all right now. Everyone will be a beneficiary of the robots.

11. There will be an explosion in the use of robots and robotics techniques in education in order to meet the increased demand to educate approximately half the population of the world that will now be connected for the first time.

12. Healthcare will rely heavily upon communication capability. Diagnosis will be by remote examination. In fact, the digital wristwatch will become more of a monitoring unit for healthcare purposes. Healthcare will become more personal as it is directed more to the DNA of the individual and not just to the average impact of medication and disease. Cancer will be cured widely with immunotherapy, the stimulation of the immune system itself to eradicate the cancer in the body. The concept of a hospital will undergo a significant change. It will no

longer be necessary to send patients to one place – the hospital - in order to provide specialized care, diagnosis, and medical treatment. The use of specialists can be achieved electronically without moving the patient to a "hospital." Hospitals will reduce in importance, and the Emergency Room will be your own home. The end result of this will be vastly improved general health at much lower cost, and increased lifespan for everyone on Earth.

13. The food supply in the world will increase dramatically through the use of robots in farming. The food produced will be more than sufficient to meet the requirements of the entire global population. In fact, there will be significant surplus. Food will be distributed just as now except that as robots take on more of the farming capability, food will be distributed by driverless trucks, loaded by robots, and unloaded by robots at the delivery destination.

14. Farming will also become a feature in regular buildings, either single-story or high rise, on the farm or in the city. In fact, such 'silo' farming might very well become a city function with robots directed in all aspects of food production. Food enhancement could be simplified in such a form of food production.

15. Food delivery via drones and robots will accelerate. This will also apply to cooked

food and meals. Roving food preparation units with robot 'cooks' will also speed up and reduce the cost of the delivery of restaurant items beyond just pizza.

16. More and more countries will adopt the Guaranteed Basic Income currently under investigation by Canada, Finland and other nations. In this process, the government provides payments that augment income to bring individual income to a guaranteed minimum. That guaranteed minimum will vary country to country but will move towards a standard equivalent amount worldwide.

17. During the next ten years, the first colonization trip to Mars will most likely occur as passenger space flight is initiated. These will commence with spectator trips into space with return to earth, as well as trips to the moon. That first trip to Mars is expected to take 80 days. With advances in technology this may be reduced substantially. One projection is to have this take less than ten days. More will be covered on this point in the chapter on space later in this book.

18. The infrastructure of the United States is in sad repair. It is estimated that at least 30% of the bridges are in hazardous condition. During the next ten years, a major program of revamping the entire infrastructure of the United States will be initiated. There will be a concerted start on repair of the

infrastructure in the developed world and its creation in the newly empowered developing world. The effect will be massive.

19. Sports programs and public entertainment will proliferate extensively as leisure time increases. This will give rise to more sports leagues and more games, both for spectators and participants.

20. The speed of the computer costing about $1,000 will approach 10^{16} cycles per second, the equivalent processing speed of the human brain. As quantum computing becomes applied more, this speed will increase.

21. Nuclear power will wane dramatically as a source of electric power. It is unlikely that any new nuclear plants will be licensed or built; and current nuclear power plants will most probably be decommissioned.

22. Social media, messages, postings, and profiles will increase significantly, at least triple, and perhaps quadruple, with a massive surge from the newly wired persons and areas of the globe.

23. Personal information about us derived from the web and our use of the web will permit complete analysis of how we think, what we prefer and admire, and our likes and dislikes. This will be used for advertising and for very personal political campaigns. Imagine the candidate addressing us individually with personal information and how the candidate will meet our hopes and expectations.

Imagine too that candidate actually speaking to us with the face controlled by computer simulation. You won't know what to believe.

24. The bureaucratic class will begin to replace the middle class.

25. The status quo will be demolished forever.

These are but a few of the major changes. The impact of technology and all the other factors outlined in the following chapters will have the tsunami effect on the world economy. The pace of these different initiatives will be different in each case. To some extent, it will depend on the economic factors as well as the desire for change in particular nations, and in particular areas of application. Government structures, legal structures, as well as corporate organizational structures will all undergo major changes. To say that the technological advances will be highly disruptive is an understatement. Not only will corporations, occupations, and machinery disappear but, just as dramatically, totally new redirection of global efforts will occur. The government will become the guardian. But new levels of taxation will occur. While the mechanical cost of government will be reduced through the use of robots, the total cost of government will dramatically increase, as it takes on more of a roll of distributing income to those unemployed or unemployable.

Entertainment of the masses will become an important element of government. Just as in the days of the Roman Empire, where the citizens were entertained to a large extent by spectacles in the forum and other locations throughout the empire, so entertainment will become an important element, especially sports, to occupy the time of the population.

Law enforcement will certainly increase in order to cope with the disruption and associated riots that will undoubtedly occur. The riots that occurred after the Gutenberg press was introduced in the 15th century are indicative of the disruption impact and resentment leading to riots against technology and the new concepts.

The military will also become an attractive source for jobs. Conflicts will become more prevalent as citizens seek to better their situations by conquest. This could be especially so with the nations that have their oil revenue shattered. Just as in the historic past, the first son of most families will most likely go into the military. However, robots will replace the soldiers in battle. The number of jobs for humans in the military might well be simply and sharply reduced.

The second son will undoubtedly follow the historic precedent of entering the ministry. There's no doubt that religion will become a growth industry as people have more and more leisure time and will seek answers to the questions of who are we, what are we, and why are we?

As a result of the increased emphasis on entertainment, the creation of motion pictures and videos will be a major growth industry. Sports events, and of course pageants of many kinds, will become major elements of life. Hence the creation of content for entertainment will be a major growth industry, as will the creation of stadia and playing fields.

It is certain that the entire world will be wired. As a result, it is certain that there will be disruptive change in many industries. The pace of development, if anything, will increase. It must be remembered that at this time there are more trained scientists and engineers in the world now that in all of history. Furthermore, all are in constant communication. This too feeds on itself to accelerate development.

It is certain that millions of jobs will disappear; even the most highly educated may have difficulty in finding jobs. It is equally certain that millions of new jobs will be created. It is equally certain that the cost of living will dramatically decrease worldwide.

It is not expected that specialties will decrease; rather it is quite likely that specialties, especially new ones, will become even more extensive. For example, there will be great demand for physicians and lawyers to become even more specialized than they are today. They would be parts of groups offering the various specialties so that teams of lawyers and teams of doctors would be available at all times electronically and support service would be provided to individuals. These teams will become the equivalent of the personal care physician, and will also give rise to the personal care lawyer. In fact, various specialty groups in just about every area such as health care, legal care, taxes, homework, education, career planning, etc. will become available to the individual.

While the size of the Smartphone will be small enough to fit within the ear people will not always want that. People will want the privacy of the typed message rather than saying it to maintain the privacy of the message if they are in a public place, surrounded by people from whom they wish to maintain privacy; or they might want to avoid any disturbance to others in a public place. In any event, there are many instances where people would not want people to place the cell phone in their ear, or be able to speak to it. Some form of keyboard will still be a necessary option. The cell phone instead of being in the ear can be in the beltline, in the pocket, or in some kind of hand held device. A projection option will allow the image to be enlarged, even projected on a wall, or condensed to appear in glasses similar, but more advanced, than the current Google Glass. Without doubt, even with all these options, it is certain that the physical size of the smartphone being used will still fit within the hand of the user, if so desired.

In any event, the Smartphone will cost significantly less that it does today. As already stated it is possible to secure a rather complex capability cell phone for less than $10 from India now. Developments are underway to have robots assemble smartphones. Currently smartphones have been designed to be an assembly of four different micro cards. Robots will assemble these in different combinations to produce smartphones with different characteristics. The expected cost will be one or two dollars, or even less. This will certainly meet the needs of the billions of persons currently without such capability.

Impact on the Economy of the United States

At the time of writing this book, the GDP of the United States is about $17 Trillion, and growing at less than 2% per year. The national debt is about $21 Trillion. In the next ten years, GDP Growth can very easily increase to 4% or more per year. If so, within the next ten years, the national debt will not increase, and the ratio of GDP to Debt will fall, possibly as low as 67%. Furthermore, the increased revenue, even with lowered tax rates, will quite probably cover increased costs of government, and even by used to reduce the national debt. While disruption will be widespread, the economic opportunities and gains will far surpass the losses. That is because the diversity of the people of the United States, and their cultures, has created an ability to grasp the new without mourning the old. This may not be true of other cultures and nations that tend to resist change of any kind. However, the wired world of the future will in itself be a major catalyst to promote change, and acceptance of change, as the benefits become evident through the medium of social networking.

Implementation Schedule

Life in the connected world will be different. It will certainly be more convenient as robots and inexpensive devices do more for you. It might very well be a much more dangerous world than we have today.

The pace in which these different initiatives will occur is different in each case. To some extent, it will depend on the economic factors as well as the desire for change within each nation or geographic area, and in particular areas of application. This will be covered in the last chapter of this book, with particular emphasis on the dramatic change in the need and use of oil, and the impact on the economic stability of the oil producing nations.

The world will never be the same again. The technological tsunami that will strike in the next ten years will produce disruptive and permanent change. Think for a moment. The whole world will be connected. Smartphones will cost pennies. Robots will be everywhere. Cars will be essentially electric. Millions of jobs will disappear forever, but millions of new jobs will be created, but not necessarily close by. Governments will begin paying its citizens a Guaranteed Annual Wage.

Oil producing nations will no longer have the great economic return they may have today with oil exports alone. Education for export, entertainment, leisure living, and resorts will join financial conglomerates in these nations to create and replace revenue lost from lower oil exports. Citizens will have much more leisure time, much better health, greater longevity, vastly improved care facilities, and the eradication of scourges such as cancer. The golden age of technology will be at hand; and so might the total disruption of the political climate of the world.

Welcome to the connected world!

CHAPTER 2
Advances in Technology:
How We Got There

Advances in technology will continue at an ever-accelerating pace.

Within the next ten years, we will have captured the capability of speeding up the computer at least a million-fold from its present speed. Whether with quantum computing, sharing facilities with cloud computing, or extended parallel computing, the computer of the next decade will be about a million times faster than current high-speed computers. Storage of data will be miniaturized to such an extent that something the size of a postage stamp would store many terabytes of data. Furthermore, the computer of the future will be miniaturized dramatically. The idea of the PC (Personal Computer) will go the way of the old mainframes. Similarly, the laptop will disappear.

If there is to be no more laptops, then what is to be the replacement? Frankly, I think it will be the large Smartphone or the small sized iPad. Both give you a reasonably sized screen, and all the capability of a laptop with something that is held in the hand. The one exception to this will be when people work a full day and do not want to use a small screen. The solution is that the equivalent screen can be projected onto a wall or onto something sitting on the desk. That something can either be flat piece of paper or an electronic screen. In any event, a larger screen can be made available one way or the other if the small screen on the cellphone or the small iPad is not desired.

There is some question about communication by voice to the smartphone. So, we have the case of apps and texting. Many people do not want to use voice in-order-to maintain privacy in a public place, such as a railroad car, and may look for other means of communicating. Certainly, means of creating texts can be accommodated through the use of a plug-in keyboard. As something to be considered for implementation at some future date, there can be silent speech. Experiments are being conducted now for the use of the echo skeleton being controlled by thought. Perhaps it is farfetched to consider the idea now of having silent speech for the brain to transmit the equivalent of words with nothing being spoken. Frankly I think it is possible, but it would be many years. But I've been fooled in the past. Whenever I've thought something was many years away it was done the same year. I don't think this will be done the same years, but perhaps instead of many years it would be less than five. On the other-hand it could be as many as twenty-five.

Nano bodies, as explained later, can be woven into the cloth so that they can be part of apparel and can be used for creating an electronic barrier for security purposes. Big question can they go in the laundry? Yes. Can they be used for anything else? Answer: Whatever Nano bodies become useful for, they can become useful if they are still woven into the cloth and the fabric of the apparel for the individual.

The machine of constant common use will be the smartphone and even it will change dramatically. The smartphone of the future may, as already mentioned, very well fit in your ear, or be part of your clothing. Communication to it will be by voice. The screen would be either sophisticated goggles, much more sophisticated than the Google glasses of today, or would be displayed on a wall or on a form or pad of paper on the desk. Indeed, the clothes we wear will be built with Nano cloth; Nano bodies will be woven into all apparel, providing a capability of communication antenna. It would also provide shielding so that all of our computer signals could no longer be copied externally. Security will be much enhanced. Some of the other developments that will occur are as follows:

Smartphones

These will decrease in cost dramatically. While the usual net cost increase for the new phones is of the order of $400 or $500, a company in India is now advertising an equivalent Smartphone for $10. Even this cost will be lower in the future. As a result, cost will no longer be a factor in the emerging world of users. While we have more Smartphones and control devices currently than there are people on Earth, not every person on Earth has a smartphone. If they did, this would triple the current smartphone population. Just think of the economic impact of that. And that is what will happen when the entire Earth is wired.

It is the Smartphone that has brought the world to the hand and fingertips of the user. Everything is available. Anyone can be connected, and messages can go back and forth in any language, with or without automatic translation. Using virtual reality or augmented reality the person or persons anywhere in the world can appear as if they are all in the same sitting room. Augmented reality is the technique whereby this is accomplished. Virtual reality, on the other hand, is associated with creating a pseudo-world, a building for example, and having you walk through the building. For you, this is a real building. A combination of virtual reality and augmented reality would be to have you appear within the images of you walking through the building.

The new Smartphones have three cameras – two in back and one in front. By having the two cameras in the back of the unit as well as one in the front it will continue to have the ability to take pictures in the front and back. Having two cameras, on the other hand, in the back of the unit will allow that second camera to be used as a tracer camera in the event you are using your basic camera for virtual reality. In that fashion your smartphone can be used for virtual reality. A helmet is not needed. Smart glasses might be required but perhaps even that will be made possible through projection.

With this combination traveling across the world to have a personal meeting will be unnecessary.

Just think of the impact. You can have a personal tête-á-tête with anyone at any time, even if the other person is on the other side of the world. You can pick a famous location. For example, you might consider having a meeting in Monte Carlo, Buckingham palace, or at the Vatican. Indeed, you might be meeting with people in all of these places simultaneously, with the decision to pick one of the locations as if you were all there together. In the future, it is conceivable that the person on the other side of you is on another planet.

My invention of the Smartphone combined all the power of the computer with all the resources of the internet with the communication capability normally associated with telephony rolled into one instrument which can be held in the hand; or in the future it might very well fit in your ear. My idea was to bring the world to the fingertips of every person in the world and beyond. I know I changed the world forever. The date of my patent filing was May 19, 1995. I demonstrated the first prototypes of the Smartphone in 1997. By 1999 I had it reduced to the current size that fits within the hand. The first prototype was a cubic foot in size. It had a touch screen just as every one of my CyberFones, for that is what I called the Smartphone.

There is no doubt that the Smartphone of the future will be an integral part of the person. In the future, everybody will be connected.

Just consider walking down the street with various companions from different parts of the world according to different situations that you are either investigating or involved with. If you have certain skills associated with, for example, weather supplies, then you could be consulting with various individuals in different parts of the world concerning weather supplies for their own use, while there will be a major resource requirement as their world develops. At this time, there are major developments leading to rapid desalination of water, which can become a major capability to be imparted to those with a need for fresh water.

Quantum Computing

Mathematicians began considering computability and the development of computing machines as early as the 1930's. Alan Turing is famous for having developed an early concept of a Turing machine. Other mathematicians have added to the concept of considering the range of numbers from zero to one. Every zero has certain variations on the atomic scale and every one has the same variations. Hence zeroes and ones can have many different states allowing calculability to proceed on many parallel paths. The result is computing speed significantly faster than the fastest supercomputer today. Hence the concept of the quantum computer dates back to the earliest consideration by mathematicians of what would constitute a computer. It is recommended that the reader consider various descriptions of this concept, which may be found on the Web.

I must apologize in advance for not attempting to simplify this concept at this point in this book. There isn't much more that could be said to simplify it other than the fact that in the normal computer we rely on two conditions, high and low voltage to give us the equivalent of a zero or a one. With quantum computing we are considering various particle levels of voltages as associated with in quantum theory. As a result, we have many conditions within a specific particle as such, instead of just two conditions for a zero and one. By having these multiple conditions at every point results in significantly increased speed of computation. Other than that simplified explanation, we would be involved in significant descriptions of quantum theory. That is beyond the scope of the book. Hence, the reader is referred to the web.

Big Data

Initially the concept of storage was to simplify the process of calculation. This was later extended to the incorporation of data files to replace physical files. Once the cost of storage became ultra-low, and once the amount of data that could be stored in a small space became ultra-high, the concept of big data became possible. This is to have vast amounts of data available for immediate manipulation by fast machines or by a number of small machines operating in parallel. For example, all of the cars leased through leasing agencies can be considered for cost of maintenance. Alternately, as has now become the famous example of Big Data, people looking up flu symptoms in a particular area can forecast an epidemic in that particular area. In other words, an inordinate number of flu symptoms look-ups in a particular zip code would signify that people there are having a problem with certain flu symptoms.

The use of Big Data has no limit. It can be used to calculate seeding cycles, maintenance cycles, maintenance requirements, sturdiness of a particular device, etc. It can also be used to identify the buying habits in particular geographic areas. The examples could go on and on. It really amounts to the unique ability to manipulate vast amounts of data, apparently not related, in order to find trend lines of importance and of use. On a more mundane level, Big Data is used by Google to establish the interest characteristics of Google users. This can then be used to pinpoint advertising messages according to interests of potential viewers of Google websites.

3D Printing

3D printing combines the cad cam process with concepts of jet printing. In the 3D printing process, material is sprayed following the computer contours as in a cad cam approach, sprayed in horizontal layers. As a result, a three-dimensional object can be "printed." This process can be used for making anything. Currently, experiments have been conducted where apartment complexes have been built, single houses, a computer, and on a day-to-day basis parts for the space platform. 3D printers with multiple nozzles can be used for spraying combinations of materials on different layers, or even in the same layer. The requirement afterwards is to coat the entire product in order to ensure that the sprayed-in materials stay in place. Various adhesives can be added to the sprayed-on material so that there is a certain adhesive quality.

3D printing will revolutionize parts inventories, making them almost unnecessary. 3D printing will also be highly instrumental in creating prototypes, speeding up the process immeasurably.

Security

With quantum computing the concept of encryption as we know it may become invalid. The speed with which anything can be handled with quantum computing will make standard encryption techniques, even with a 256 or 512-bit key arrangement capable of being cracked within a matter of minutes. Hence the security systems of the future will defer mainly to hardware. With regard to personal characteristics of the user, these will go more to the actual persona of the individual, iris scan, facial scan, fingerprints, and any other measurable entity linked directly to the body of the user. This at least will prevent unauthorized users from entering or penetrating any system.

Just a footnote here associated with my personal belief that many of the publicly acknowledged penetrations of major government databases I think were perpetrated by hardware back doors and not necessarily by hackers. For that reason, it becomes rather important to maintain any kind of security to have complete control from beginning to end of all hardware components and chips in any system.

An effective security system can only exist if administrative controls are rigidly adhered to. It is the nonuse of such controls that would make penetration of the most secure system possible.

For example, it is standard practice to present the improper use of highly secure equipment by operators who are not cleared for the use of such equipment. Not doing this would become an open invitation to fraud, and significant loss, if not destruction of the system.

For example, in the creation of many funds transfer systems, the operators of these systems capable of transmitting control of vast sums of money are normally separated into cages. Access to these cages, and resulting access to the computers, is highly restricted only to those cleared for such activity.

Another administrative situation is to have three people involved in the transfer of money - one to enter the original data and a second person to reenter the data without knowledge of what has been entered by the first person. This second person would also have the ability to edit or change. And by the way, any time any change is made to any record, an audit trail is made of that in terms of the new value, the old value, and who did it and why.

The third person would have the ability to release existing wire transfers, but would not have the ability to enter or change. Such a person would, of course, have the ability to cancel an existing wire.

Currently, most wire transfer systems give third party, or users, the ability to enter all the details associated with a wire transfer. These can be standard wired, where only the amount has to be entered for it to become effective. Having created many of these systems, on a first-hand basis I can say explicitly that this is very dangerous. Unless there are adequate controls in terms of how these are set up and monitored, creating this ability should be discouraged. If it is allowed, then there should be adequate safe guards to ensure that the party sending the wire has the full responsibility for any malfunction or theft associated with this process.

In the future, there will be many forms of hardware, software, and other access control procedures. All of these might appear to be restrictive, but should be instituted, and controlled in terms of performance if security is to be maintained. Remember that the amount of money transferred in any single day is often multiples of the total asset value of the bank. With regard to third party transfers, it should also be remembered that quite often these transfers are also a significant percentage, if not multiples, of the total asset value of that company. As a matter fact, more importantly, even if there are adequate controls, it should be remembered that every attempt would be made by thieves to circumvent all of these controls. Recently there was a description of how thieves attempted to divert some $950 million from the world within the SWIFT system (Society for Worldwide Interbank Financial Telecommunication - a non-profit entity in Brussels) used for international funds transfer. As a matter of fact, the thieves did get away with $95 million. And this was done over a weekend. By combining the weekend with the international dateline, in other words by using a bank on the other side of the world, it was possible to use two weekends, and hence have a total of three days during which it would be possible to have the stolen funds totally disappear.

Note that there is no limit to the amount that can be diverted or stolen. There is further no way that the trail can lead to a successful recoup. Quite often a trail will terminate as a dead end. For example, many of the casinos in Philippines and Macau cannot be penetrated in terms of the disposition of funds that enter their accounts. The same is true of many situations in different locations around the world. Casinos, of course, are notorious. But so are many of the financial havens, most especially in some of the smaller countries of the world.

I would hazard a guess, if not a wager, that millions are stolen each day. Beware!

Healthcare

During the next ten years, significant strides will be made in healthcare. This will move from the traditional concept of a patient undergoing many different tests with many different doctors, all directed by one primary care physician. First of all, the patient will not be required to travel to these different locations. In many cases the encounter, including a detailed examination, will do electronically using virtual and applied reality. The instrument of use will obviously be the Smartphone. There will be certain attachments that can be added or plugged into the Smartphone, which will include a stethoscope, an examination item for sore throats, ears, etc. A blood pressure device will also be attachable to the Smartphone. Finally, a sensing glove will also be attached which will to give the physician at a distance the ability to palpitate the body of the patient. Hence, to a large extent, diagnosis, and even certain treatments, will be possible at a distance. The result of this will be the vastly increased ability of physicians to handle significant numbers of patients. Furthermore, such examinations will be highly personal directed to the patient. For example, by incorporating DNA for each patient, it will be possible to determine the impact of different medications on the specific person according to the DNA reaction. Medicine, diagnosis, and treatment will be heavily personal, and not depend upon averages. In that fashion, many adverse reactions normally recorded will not occur.

Perhaps of major impact will be cures for many diseases, including cancer.

Great strides have been made with regard to new concepts of curing cancer. These are associated with boosting of the immune system so that the immune system itself eliminates the cancer cells. In fact, properly administered, the boosted immune system would make cancer not only disappear, but impossible to reappear.

The medicine of the future will be associated more with prevention than with cure. This has to do with every disease known.

Big Data, of course, will also be a major asset in determining the factors that lead to various diseases.

We will certainly establish new diseases as every person on Earth becomes connected and as we venture into outer space and interplanetary travel.

Solar Power

Solar technology is improving every day. The efficiency of solar panels currently has advanced dramatically. In the next ten years, it is certain that this will triple. The impact of this will be to make solar panels economically feasible everywhere in the world for every household, and for industrial use. Furthermore, the industrial solar panels will come in rolls that can be merely taped or nailed to a surface. Expensive and long-term installation will no longer be necessary.

With the extensive deployment of solar power, there will also be a deployment of stand-by batteries to absorb excess power not being used at the time providing power when there is no sun, as at night, or during storms. Significant strides are being made in the development of zinc-ion batteries that will provide such capability without the hazard associated with lithium-ion batteries currently in such use. In fact, at this time, during the writing of this book, contracts are being negotiated by major electric utilities for large-scale stand-by zinc-ion batteries for wind, solar, and tidal systems. Major industrial use of power generation will come, in the future, from tides, wind, but mainly from solar power. Electric power generated by oil-fired units will decrease during the decade ahead, as will atomic power. Nuclear power will cease to be a factor in the electric grid with phase out beginning during that ten-year period.

There are some countries and locations that are blessed with the ability to use hydroelectric power. Wherever that is it available, it will certainly be the number one asset for the generation of electric power.

Without doubt, during the next ten years, the use of oil products, and nuclear power, will begin to wane in their use to generate power in the electric grid. This will have a major impact on the economy of the world, as well as the cost of living. It is estimated that the power will cost pennies with these conversions to renewable sources.

More Changes

The 3D printer will revolutionize innovation and maintenance. It will no longer be necessary to have a whole storeroom of spare parts. When a part breaks the 3D printer can print a new one. All that is required are the design characteristics that would be maintained in a data warehouse.

These advances in technology in the next ten years will provide tremendous capability for advancing the economy of the people of the world. Which raises a point. Currently we have jobs in industries going to different countries in the world. In a sense, we have globalization of production. We have to a large extent globalization of financial systems as well. The Export Import Bank, the World Bank, and the close association of the reserve banks of each of the nations of the world are indicative of the fact that the global economy of the world even though segregated within nations, does operate to a large extent as one. The day will come when that will be the case. It may not occur within the next ten years but certainly within the next twenty-five. Within the next ten years many political separations of people will begin to disappear. The most famous attempt at this was the European Union. While each country maintained its own culture and language, there were moves to consolidate this. Many Europeans spoke multiple languages. Just think, the diversity of the various states of the American Union are probably as different as the differences that are traditional within the nations of Europe. And yet we have this vast territory in the United States as one nation. The same is true of Canada. Can we look to the day when the nations of the world will realize that their future lies in their cooperating with each other rather than erecting barriers or walls that will separate them?

Let's consider the cost of living. This will drop dramatically in the next ten years. With the advent of solar power that is effective and inexpensive, the cost of power in the individual home will be cut to less than a fifth of what it is today. The 3D printer will make home building less expensive than it is today.

Robots will perform many of the chores in the home and in industry, significantly reducing the cost of living and the cost of goods.

It can be conjectured that the total cost of living will approach twenty percent of what it is today.

While the cost of living will certainly go down, there can be personal hardship in the absence of a job. Hence, subsistence will be by government grant as more and more governments introduce the concept of a minimum guaranteed income provided by the government.

The problem will be that there will be significant leisure time. As a result, governments will be required to provide festivals and entertainment to occupy the time of the people so they will not riot or engage in other activity detrimental to the general public welfare. Hence, sports events will increase, as will circuses, and to large extent public festivals. This will require the erection of many stadia as well the creation of new academies for sports and entertainment. This will provide income as well as employment.

The major advances in technology during the next ten years will be associated with virtual reality, applied reality, education, artificial intelligence, and space. All of these will be considered in detail in the appropriate chapters associated with those subjects.

At this point, it should be stated, however, that virtual and applied reality will make it possible to have meetings with different people all over the world, and in the future from space as well, all meeting in a virtual setting as if they were in the same room. In addition, it will be possible for the examination of buildings without having to travel to the building. Hence space can be examined for rental or purchase purposes, or repair purposes, or whatever, without actually having traveled to that location. As already indicated, technology will be vital in healthcare, and will be significant in research once again for the ability to have on-the-spot meetings with people all over the world, all at the same time, and the same place.

The impact on education will be significant. It could be stated that without these capabilities it would be virtually impossible to meet the educational challenges and requirements of the next ten years and afterwards because the entire world will be wired and joined for the first time ever.

Artificial intelligence will be the major essence of many of these newer developments and hence require extensive treatment in this book. The same is true of the developments with regard to space.

CHAPTER 3
Your Life Ten Years from Now
Our Disappearing World

Your current job may disappear in the very near future. You should be alarmed, but not unnecessarily so. If you have the drive and the desire and the training, you will take on one of the new jobs that will emerge. If you are completely unskilled, you may have a severe problem in finding a job.

Just consider what has happened to many jobs in industries during the past seventy years. Typewriters have disappeared and been replaced by word processing. Word processing in turn has been displaced by voice input. Hence a machine process has now replaced the stenographers who took shorthand and dictation. Clerical jobs have almost all been eliminated. At the same time, thousands of new jobs have been created. But most of these require special skills.

Hundreds of thousands of jobs have been created in the information industries centered on the computer. All aspects of expanding the use of space were made possible not just with the invention of rockets but also more importantly with the invention of the computer which made the control of space flight possible. I well remember the look of startled disbelief on the face of my research director in 1952 or so when I suggested to him that the space vehicle of the future would have a computer on board. I should have made that a plural. He, by the way, was a very knowledgeable man. He was a Fellow of three of the most prominent aeronautical societies in the world. After I graduated we socialized a little bit and one time after he got his third fellowship we went to the movies and I addressed him as Sir Triple Fellow.

What was my approach to space? I always wanted to go to the moon from the time I was a little boy. I always wanted to fly. My brother and I, as I related in my autobiography, haunted the airports around Toronto getting to know the pilots, many of them Aces from the First World War. For my doctorate, I was assigned the task of determining the heat transfer characteristics of space vehicles as they reentered the atmosphere of the earth. Nothing was known at the time, the early 1950's. It was necessary for me to use a computer to solve the equations that I could not crack in closed form. Hence in the early 50's I became an expert in two of the great development fields of the next seventy years - aerospace and computer science. I have straddled both fields ever since.

I saw whole new industries created and I saw the demise of old industries. Sitting over dinner with our four sons one night we got into a discussion with our son Paul, who is a noted expert in privacy and security. I find this rather ironic since Paul, as a lawyer knows as much technology as the other three boys, all of whom are considered to be experts in technology. Anyway, Paul asked what my current book was. I told him. He suggested that I might want to do a book on "Rest in Peace: Your Job." The thought struck me hard. Yes, many of the jobs of the present will be gone. The politicians who are continually claiming that they will create thousands of jobs and bring back the jobs that fled overseas are all wrong. The jobs that fled overseas will now disappear even there. Simple jobs will all disappear in time. They will be gone, gone forever. Manual jobs that can be done by people with little or no training can be totally replaced by robots. There are no ifs, ands, or buts about that. Any job that is repetitive and can be performed by people with no skills whatsoever can and will be totally eliminated. The robots and computers of the future can even replace many of the jobs that require great skills. But, thankfully, or hopefully, they will create vast new hordes of jobs that would not exist without such capability. With "fire in the belly", a vast number of new jobs will be created, creating new industries, new opportunities, and meeting the challenges of the totally wired world. Rather than "Rest in Peace", it would be an explosion of opportunity and challenge. Paul agreed. That's how we came up with the final book title.

One of the nicest or greatest or happiest in a sense of these new jobs or new capabilities is something I discussed in the 1960's with a great friend of mine, an orthopedic surgeon who became a physical medicine specialist as well. He had spent most of his professional life helping people through exercise and through manipulation of the affected parts of the body. He found that through exercise, manipulation and injection he was doing a better job without surgery.

We often discussed the case of a paraplegic. To what extent was a paraplegic afflicted by an inability of the signals in the brain to manipulate the muscles to which the brain signal was directed? We never could solve that problem. But we did come up with the idea of encasing the body with a type of structure or skeleton that would move the different parts of the body. I was convinced that the computer, even though this was over fifty years ago, would become the small, compact, handheld device it is today. My concept was that a computer would be carried by a paraplegic who would have this external super skin which would be commanded by the computer so that the paraplegic would then be able to walk and to move his or her arms or legs. That of course is what we are doing today.

Universal Wi-Fi

The most important aspect of the future is that the modern satellites and the modern communication capability will make Wi-Fi universally available. In the few short years from the date of this book the entire world will be covered by access to the internet. Just think what that means. Currently we have less than half of the population of the earth capable of using the internet. Just imagine the impact of broadening this to the entire population of the world. Just think of the jobs that will be opened up requiring the training or supply of technology and technology instruments to more than half the population of the world. You might say that they wouldn't be able to afford anything. Well right now we have lowered the cost of the iPhone from $700 to some of the older models down to $200 or $300. There is a firm in India, Ringing Bells Pvt. Ltd., that currently is advertising smartphones for less than $10. These are available now. They compare quite well with various supermodels of the Smartphone available at many times that cost.

More will be said about these developments in the later chapters of this book. The first part of the book describes the disappearance of the jobs and gives you some idea as to how large a component in terms of numbers that is. The second part of the book is concerned with the promise and perils of a true extension of technology to everybody. As stated earlier, the cell phone of the future may very well fit inside your ear. The keyboard will be replaced by your voice. The display screen will be replaced either by glasses you will wear or will be transmitted to display on a wall nearby or on a piece of paper on your desk. And so, your old job disappears and you welcome the new job, assuming you get one!

Throughout the world, we are approaching a technology tsunami in the next ten years that will sweep everything before it. Life will not be the same. This might appear to be a drastic or scary prognosis but the seeds are already in place. Just think, what has been the impact to date of the digital computer-radar, satellites, space flight, the internet, the Smartphone, and on and on? As a matter of fact, the start of the predicted tsunami was in February of 1946 when the first digital computer, the ENIAC, was announced and displayed in public demonstrations. The year 2016 marked the 70th anniversary of the birth of the digital computer age. Much has happened in those 70 years. As a parallel invention, a few years earlier, radar and three-point navigation systems were set into development. The jet engine also saw its birth in the 1930's leading to its use towards the end of the Second World War. As a matter of fact, it was the development of the jet engine projecting global travel that led Marshall McLuhan to formulate his concept of the global village. This concept was elaborated on in books that he wrote in the 1950's. I was fortunate to have been at the University of Toronto when Marshall McLuhan was there and I often attended his lectures. He was struggling to formulate this concept of the Global Village in lectures he delivered that I attended in the late 1940's and early 1950's. His additional message, "the medium is the message" and "the message is the medium", has come to pass but in a different way than he envisaged. He was thinking of print. The concept of electronic media was just a dim thought. It is the electronic medium that is now the deliverer of the message. So, we have the start, the seeds of the tsunami. The birth of the jet, the birth of electronic navigation, and then came the Internet, the Smartphone and social media. These three elements were catalysts that finally set in motion the tsunami, which will sweep all before it in the next ten years.

With all the changes that technology will bring about during the next ten years, two things will affect your life significantly. The first is that it will be easier to live and secondly that your job may disappear. Jobs will be an essential element in the future. If somebody has a job, that person will be blessed. As said earlier, many of the jobs as we recognize them today will be totally gone. They will not be shifted overseas but they will be gone. It must be recognized that with the changes that will occur, millions of jobs throughout the world will absolutely disappear – forever!

We are in the midst of a presidential campaign in the United States as I write this. Both candidates are promising to create jobs. One candidate is promising to bring back the jobs that left the country to go overseas for lower labor rates. Sorry to say candidates, but you're both wrong. No matter what you do, you will not be able to bring back the old jobs. But millions of new jobs will be necessary. The key is to create these new industries and new jobs that will be necessary as-a-result of the technology tsunami.

Let me tell you what these will be. But first, let me justify my assertion that many current jobs will disappear forever. As an example, consider buggy whips. Typewriters. Carriages. Hot water bottles. Camera film. 78-rpm records. And on and on. Millions of jobs gone. But consider the millions of jobs in the information industry. Consider the wealth generated by the new technologies. So, it will be in the technology tsunami of the next decade.

Let me tell you how. Technology in the form of robots will replace many of the jobs that we currently take for granted. At the current time, nursing is a growth industry. Nursing jobs are increasing annually. As the robots take hold, these jobs will start decreasing and gradually become a smaller and smaller part of the workforce.

You might consider programming, a very highly skilled occupation directed towards creating the programs that drive technology. The idea of creating automatic programming systems has existed from the very beginning of the computer age. The first so-called compiler, a program that creates the running code that drives the computer on a particular application, were first conceived of as early as 1948. These have improved over the years to where we are rapidly approaching what are called natural language compilers. That means that you will speak into the computer and the computer will generate the code to produce the system that you described. Hence the hordes of programmers associated with creating the code will no longer be necessary or needed. Highly skilled systems designers will be in great demand, but programmers will not.

Under various considerations of the changes to occur in the next ten years, I have indicated that a great deal of leisure time will be available. I have further indicated that entertainment sports programs will become important elements in coping with the increased leisure time of people. This will require some form of babysitting by government agencies so that the population will not become restless and become engaged in riots. One of the problems in today's environment before the tsunami of all of these changes strikes is the amount of leisure time people on welfare have. Such individuals are prone today to riot and demonstrate since they have plenty of time to do it. The amount of leisure time as such will increase as the tsunami of these changes occurs. This will be specially pronounced when people are unemployed to a greater extent than they are today because their jobs will disappear. It will be necessary to take up their time. This was similar to the problem in the days of the Roman Empire when games and entertainments were created by the emperor to keep the people occupied and happy. The same might very well be necessary in the situation now. Sports programs will be extensive. This will in turn create jobs for the athletes. The cost of providing such entertainment and sport spectacles might very well be a tax benefit. Taxes, as such, will certainly go up. But then again they might not. If many costs in the government are offset because they are no longer necessary, then additional expenditures can be incurred without having to raise taxes. Furthermore, the new industries could very

44

well increase the Gross Domestic Profit (GDP). This would then generate more income for the government permitting more serious and perhaps even lowering tax rates. This will require significant study and certainly would be the appropriate and hoped for solution. That would provide the extra or additional revenue that will be required to underwrite the sports programs, and to write the Minimum Guaranteed Wage to be provided by the government.

Any repetitive job can be replaced by robots. This has already occurred, especially in auto manufacturing. On the other hand are there current robotic jobs that we have found that work better via that provided by unskilled human labor touch to do the task? This is the case where artificial intelligence can be coupled with robotic capability to provide companion capability for those who are sick and infirmed.

Such situations will certainly be found to a greater extent as robots are deployed for various activates in the very near future.

For people who are technology phobic, have difficulty grasping technology concepts or lack the skills for technology, where would they find a career in the future job market? That would be difficult. But such is always the case. There is a certain element of the public that cannot be trained in technological pursuits. These are the kind of people who are suitable for nursing or for companion jobs. They would certainly be very suitable to be teachers. There is no doubt in the mind of the writer that jobs will be available for people who are technologically phobic. In fact, the number of jobs for people with no technology capability will be large in number. The difficulty, as I foresee it, will be to interest people in such jobs. My feeling is that most people will seek technology-supported jobs since they will be using technology to a large extent in their own personal daily lives. Hence to ask these people or such a person to take a job that has no technological requirement or capability would be counterproductive.

It might be said that people who are skilled in organizational repetitive tasks, which can be replaced by robots, will have great difficulty in finding employment in the future. If they are not athletic or artistic they might have even greater difficulty.

This is a rather unique question, and I don't think that this can be answered now. I think there are as many opinions as there are people. My opinion is that this will not be a problem. My opinion is that it will be difficult to fill all those jobs that have no technology requirement whatsoever. Individuals in the future will probably shy away from such activities since it will not, in their opinion, use their skills.

Let's wait and see. In any event, I think this will be a small factor in the great turmoil that will be going on anyway.

This raises question of why turmoil will occur after the jobs disappear and when there is more leisure time. That's just it. People with leisure time usually fill it with mischief. Hence it becomes important to make sure that the amount of leisure time people have is kept to a minimum. If they have nothing to do, by and large quite often they do things that you do not want them to do.

They would also become, of course, susceptible to the blandishments of rabble rousers who for their own purposes and objectives would be stimulating discord, riots, and seeking to undo the benefits of the technological advancement. This occurred to a large extent with such implementations as the printing press. There were tremendous riots with the introduction of the Gutenberg press when it was introduced in the 15th century. In fact, and going back through history, in every case where great technological implementation occurred, there was significant discord at the beginning and significant resistance in change.

Let us take another highly skilled occupation – surgeon. Currently robotic machines controlled by a surgeon are performing very complex operations. In some cases, these operations are performed better in terms of delicacy by the robots than is possible with the human hand. While the skill of the surgeon will always be required to ensure that the robot is performing properly, as this capability improves with time, the need for a surgeon in many cases will be rapidly reduced.

These are the unexpected reductions in labor force. The expected are clerical and sales jobs. The agent that sells you insurance will be replaced by a computer program operated by a robot. Personal care will be performed by robots.

Attendance to the elderly and the infirmed will also be materially improved with the use of robots. Nursing homes may become an element of the past as robots can provide all of the assistance necessary in the home. And this goes on and on. In the various chapters of this book I will attempt to outline the expected progression of robots, drones, virtual reality, applied reality, 3D printing, artificial intelligence and all the major changes in what we call computers. Just for the moment let me say that laptops to a large extent will disappear. The Smartphone will continue in two forms, certainly in the form we have today but also in a form of a unit that will remain in your pocket or possibly in your ear. Instead of typing one letter at a time, the voice entry, which we have today in a rudimentary fashion, will become highly sophisticated and will become the norm of communication with your Smartphone. The Smartphone in turn will be your total control unit, operating not only as we know it today, but also in controlling the car that is driving under your control but not with you at the wheel, accepting deliveries from drones and from robots, controlling the robots in our homes that are providing the basic services of support including cleaning, cooking, making the beds, washing our clothes, caring for children, etc. The Smartphone, then, will become the central control unit by which you will interface with the highly technical world that will develop during the next ten years.

But be aware. Your job may be at risk.

In today's environment, 2016, there are approximately 2.5 billion Smartphones in use. The world's population is close to 7 billion. Hence there are billions of people who are not connected. Imagine the impact and the demand for services if all 7 billion were connected. The likelihood is high that that will be so in the next ten years. Consider where we are with the use of the Smartphone now, and where we are going to be in the next ten years with regard to services and impact on every aspect of our daily lives. The impact at this time, even in the infancy of the Smartphone, is profound. Over 1.5 billion are connected with Facebook alone. Educational programs utilizing the power of the Smartphone and tablets are proliferating. As yet, this impact is small. It is expected to grow exponentially.

Online medicine, especially with regard to diagnosis, if not treatment, utilizing the internet and a Smartphone or smart tablet is beginning to be accepted and in use. Within ten years this could become widespread, especially if the Smartphone population is extended to cover most if not all of the entire global population.

The computer was born out of a desire to calculate. The first target was artillery trajectories. That was the stated objective for the project authorized by the U.S. Army in 1943. ENIAC was born from that funding but not until the war had ended. The calculation capability of the computer continued. But soon the ability of the computer to handle large amounts of data became evident. Clerical tasks became the target for computers, justifying the development of the computer as a product. The need for cost savings plus normal technological development led to increased computing speed, miniaturization and random access storage, and the next wave of the technology revolution took place. This brings us to the sixties. The next cycle was to combine the computer with the needs and dreams of the space scientists.

Recall that the computer gave rise to space flight. Space flight made it possible to have communication satellites. Space flight also made it possible to have navigation satellites. The combination of satellites for communication and the miniaturization of the computer gave rise to the possibility of having hand-held telephony that in turn was combined with the power of the computer in the creation of the Smartphone. So now we have the Smartphone, satellites that now have changed from costing hundreds of millions to costing a few thousand, from tons of weight to less than ten pounds.

The stage is set for the technical tsunami of the next decade.

Robots will deliver your groceries. Your house will be cleaned by a robot. Your house will be built by a robot or by a 3D printer. Your wristwatch will be diagnosing your ills continually advising you what to do if you have trouble. You will buy insurance from a computer system. The checkout clerk at a supermarket will be replaced by a robot. Farming will be performed by robots.

Hang on for this tsunami. Head for the high ground, or you will be swept away. The chapters of this book are intended to point you to the high ground.

Robots, drones, 3D printing, virtual reality, augmented reality, artificial intelligence, the Cloud, Big Data – a whole new lexicon of terms. The final input of the Computer Revolution unleashed in 1946. Many jobs will disappear; but many new ones will be created. The next chapters of this book will describe how the change will come about, estimate how some jobs will be affected and then point to the new opportunities that will be created. The main point is that the world will never be the same again. Your old job disappears and you welcome the new job, if any.

Most of the new jobs created will be associated with four things. The first of these will be training of those who are newly connected. That will amount to about three or four billion people. These are people living in what has been, up to now, the most inaccessible locations in the world. One such location is Siberia. The vast reaches of Siberia, populated sparsely with people, and prior to the demise of the Soviet Union, also populated with many prison colonies. These remote areas are replete with natural resources, or require significant development. I remember my trip to Moscow in 2000 when I was the guest of the President of the Kurchatov Institute. That institute is a combination of the Los Alamos and MIT in the United States. It is very prestigious. It is considered that the president is the fourth or fifth most powerful man in Russia. In any event, this man, Dr. Eugenio Federov had, at one time, been in charge of vast sections of Siberia. He told me he had experimented with the use of barrage balloons raised to a thousand feet to provide transmission capability for cellphones and the internet. I thought this way a rather unique and ingenious solution and worked with him on the idea of having a vast number of barrage balloons. I soon rejected that and told him that we really have to rely on satellites, and it was my hope to provide plasma type propulsion for such satellites. While I envisioned the small satellites that have developed in the last five years, I did not envision the rapid development and conceptions that occurred. Social networking has been the catalyst.

In the year 2000 we were still suffering from the financial disaster of the Iridium satellites. At a cost of hundreds of millions of dollars, the Iridium satellites were deployed in hope that they would return that cost through various licensing arrangements. These arrangements were so costly that they did not come to fruition and as a result the entire Iridium network went into bankruptcy. It was rescued from bankruptcy at a fraction of its original cost which then made it economically feasible. A lot of people lost a lot of money before that became practical. As will be covered in a later chapter on space, it is now possible and economically feasible to have the cellular satellites, the small ones, weighing at most ten pounds crisscrossing every fraction of the near-Earth orbit that will provide communications in all parts of the world at all times.

The second aspect that will create vast numbers of jobs will be the necessary efforts building infrastructure where it doesn't exist, or repairing it where it has to be upgraded, as in our own country.

Peace keeping will also be a major job area for the future as vast numbers of people are either unemployed, or were never employed but are suddenly connected. This will result in significant turmoil.

And finally, as already indicated, it is expected that religious groups will proliferate and more and more people will turn to religion and asking the perennial question of how am I, why am I, and who am I?

The problem is obvious. Will the new jobs be generated fast enough to meet the needs of the hordes of people whose lives and income are disrupted by technology as their jobs disappear? While millions of new jobs will certainly be created, it is just as certain that millions of current jobs will be eliminated. I will hazard to guess that we will be eliminating jobs faster than we will be creating ones. Hence life during the next ten years will be somewhat complicated. For some it will create totally unexpected prosperity, and for others, including many with highly advance degrees, the eliminating or curtailment or reduction of their current employment and income.

The challenges will be many. On the whole, just as with all technical developments in the past, the first wave will create a certain amount of misery and dislocation, very rapidly superseded by unbelievable prosperity. That prosperity will be generated by connecting the entire world. Just think about it - adding another four billion people as potential users of the products of the world. The positive results will be tremendous. But so will the task of handling the disruptions to the status quo. Ultimately it will all be positive.

CHAPTER 4
Robots
Our New Companions

One or more robots will soon be a fixture in your life – at home, at work, and all around you. Robots in one form or another, whether they proceed on land or air, will be a universal element in the world.

We will consider a robot as any machine capable of carrying out a series of actions automatically or under direction of a controller, whether directly or via wireless linkages. As a result, we can consider a robot as any machine whether one that flies, travels across the ground, or performs a series of operations within the home, office, industrial plant, or anywhere in the world or in space. The control element of a robot can be pre-established with a fixed program imbedded in the robot, or can be set to accept changes from external command, via a control stick or electrical impulses carried by a wire or received wirelessly. The internal program of a robot can be one that is never changed, or one that can be self-modified as the robot learns more and more. Hence a robot can be completely generalized in terms of what application it is directed to accomplish, and where.

Robots can be guided by a rigid program that is an expert system designed for a specific purpose and use; or can be imbued with artificial intelligence allowing varied response and learning ability. Such robots with or without artificial intelligence are the wave of the future.

The concept of a robot is not new. Robots have been a factor in manufacturing for centuries. Automatic looms for making of cloth are one of the first robots in history. In our day, automatic pilots for aircraft come quickly to mind. They have developed from the rudimentary systems of the past to the sophisticated wireless system that allows a controller-pilot to fly a drone thousands of miles away. At a shorter range, toy drones and race cars can be controlled with a hand-held wireless unit. Remote control was also used during the Second World War. Control can also be exercised by direct connection.

The classic robot from the past has been the mouse, which finds its way through a labyrinth. That is a very simplistic design with the mouse backing out and finding another path whenever it encounters a blockage. A variation of that is the home vacuum cleaner, which is a flat device let loose in a room while it proceeds to vacuum the entire area. Whenever it bumps into an obstacle such as a sofa leg, it backs around it and proceeds. Devices of that nature are adaptable, but are not learning. They do not learn where all the obstacles are in a room. If the program is modified to take into account every obstacle that is encountered, and then regulating the path of the robot to avoid these obstacles, then it can be considered that the robot has a learning program.

In the past, some examples of robots are the toy race cars that are controlled with a handheld device. More sophisticated but yet of the same principle are the wrecking robots that seem to be the preferred game of engineering students in an engineering faculty such as MIT. It was Churchill who recommended the idea of a torpedo under the control of the wire from the submarine.

In today's environment computers are ultra-small. They can fit quite readily in a small space and require very little power. Hence, we can have controllable robots regulated by a computer operating under a preprogrammed set of instructions or modified by wireless command. The drones are probably the most famous or infamous of these. Toy drones can be purchased inexpensively and have now become a hazard to aircraft and must be controlled. In a military environment, we are all well aware of drones that are used to pinpoint enemy combatants or terrorist leaders and then can be attacked by the drone firing missiles. Such drones of course rely on GPS to find or determine their exact position.

Ground drones are being considered as delivery vehicles by Amazon. They are also considering flying drones. Amazon and the government of UK are currently conducting trials with regard to drone delivery by air and by ground.

A difficulty of controlling drones is to ensure that they do not interfere with commercially scheduled aircraft. This becomes a relatively simple matter of keeping the flying drone out of the airspace set aside for commercial aircraft. The difficulty lies with regard to noncommercial aircraft. The amateur pilot who flies at low altitude, normally below 6,000 feet, will now have to be restricted to certain areas in order to avoid contact with flying drones.

There are various conjectures as to whether Amazon will put UPS and FedEx out of business. However, it should be remembered that if drones will be useful and helpful for Amazon they would also be helpful and useful for UPS and FedEx.

Drones are a form of robot.

Robots at Work

In the next ten years, we will have significant application of the development of robots for warfare. In the home, however, the greatest application of robots will occur. Robots will become companions, servants and handymen around the home. They will cook, clean, shop and repair. Drones will also be able to communicate with people and actually carry on a conversation since they can be equipped not only with a computer to direct their activity, but also with concepts of artificial intelligence which will allow them to respond intelligently to requests. In commerce, the robots will perform secretarial work, maintenance, cleaning and delivery. In the hospital, robots can be used for nursing care, performance of examination, taking of vital signs, delivery of food, preparation of food, and even surgery. At the present time, there are surgical robots that are under the control of the surgeon but conceivably there are many procedures that can be preprogrammed so that the robot can complete the operation from beginning to end with the surgeon standing by to take over in the event of a malfunction. If we consider this to be extreme, consider the fact that the robots called automatic pilots have been flying airplanes with hundreds of people on board automatically. As a matter of fact, these automatic pilots can actually land a plane, normally better than a human pilot.

In schools the robot can be quite useful in working steadily with a slower paced student. Conceivably every special needs student can be accompanied by a personal robot to take care of all of their requirements. A human teacher or controller can thus handle many robots.

Where will this take us? Just consider for a moment. Robots will generate an entirely new industry, the home companion. Just as the automatic washing machine, dishwashing machine, and microwave have become common devices in the home, so too will the home robot take over many of the tedious chores of home life. The robot would have the dinner prepared and on the table when you arrive home from work. The robot will cut the grass, water the flowers, pick the flowers for the table, and ensure that appliances in the home are in working order. As a matter of fact, it is the robot that will be using the appliances and not the homemaker. In a sense then, this is a replacement of "housework" with a robot. What a gift for the harried housewife.

But it is the replacement robot that becomes a threat to jobs. Just think of how much of the workday of a nurse can be replaced by a robot. Twenty five percent? Fifty percent? Seventy five percent? Certainly, not one percent. Consider fifty percent. If it is fifty percent, then fifty percent of the nurses will no longer be needed. There are currently approximately three million nurses employed in the United States. If fifty percent of them are no longer needed, that would result in a loss of 1.5 million jobs.

The Delivery Robot

There is no doubt that the Amazon experiment with the government of the United Kingdom will be successful. Robots will be used for delivery, whether it is ground robots or drones. Let us assume that half of all of the deliveries made currently will be replaced by robot delivery. This of course would include postal workers. If a robot can deliver a book from Amazon, then it can certainly deliver a letter from the postal service. There are currently approximately 600,000 postal workers in the United States. If 75% of them are replaced by robots this would eliminate 450,000 jobs. There are currently approximately 700,000 persons involved in the delivery of packages for such organizations and UPS and FedEx. If robots replace 50%, then that will eliminate 350,000 jobs.

It doesn't take long to realize that robots will probably eliminate fifty percent of the routine jobs in manufacturing, delivery, home attendants, maid service, nursing services, and even teaching. What an impact!

Robots in Healthcare

The exoskeleton developments are under way and during the next decade should become prominent in their use with quadriplegics and other paraplegics.

A robot can obviously be used in nursing care in taking vital signs, in maintaining records of the patient's progress, etc. Needless to say, robots can be extremely useful in the administration of medications, and in handling such patient care requirements as bed making, cleaning, preparation and delivery of food, etc.

The DaVinci is a general-purpose surgical instrument. The surgeon sits about five feet away from the patient and manipulates various instruments including the light, camera, as well as surgical instruments. Instead of the usual major incision, the DaVinci operates through five small apertures.

The DaVinci has been used extensively in coronary operations and in prostate operations. Approximately 1,500 prostate operations have been completed at this time using the DaVinci. In a comparison with normal procedures for the prostate, the morbidity is the same. The major difference is that the hospital time, and the recovery time, is significantly reduced using the DaVinci.

The same is true in the coronary operations, most especially the valve replacements.

There seems to be no limit to the type of operation suitable for the DaVinci. Additional developments are under way at this time where the number of incisions will be significantly reduced. There is one robot surgical system that will do everything through one small incision with a general-purpose tool to perform various functions normally requiring more than one minor incision.

The future of surgical robots is very promising. It will not replace the surgeon, but will significantly reduce the recovery time after surgery, and significantly reduce the pain as well.

A variation of the surgical instrument is the development of the exoskeleton. This is something that is controlled by the human brain through an incision into the brain connecting the electrodes to the exoskeleton frame. This is something that fits around the body or around the limb for use with paraplegics. Through the stimulus of the brain the machine will move the limb as desired. In that fashion, even quadriplegics will be able to have mobility.

Transportation

We have already alluded to the use of the robot called the automatic pilot in flying of aircraft.

We've also alluded to the fact that aircraft can be flown by an external pilot thousands of miles separated from the location of the aircraft. These can be the drones as currently employed by the military to seek out terrorists and to destroy them with a rocket. Less commonly known are the drones flown by pilots remote from the aircraft, and these are full sized aircraft, used for target practice by the military. There is no reason whatsoever that cargo carrying drones could not be used by transportation companies.

Amazon is in current studies of the use of drones to deliver packages.

A more interesting drone, really a black box resident within the machine, will be the use of driverless automobiles. At the time of the writing of this book, Volvo is delivering a number of cars to Pittsburgh to be used by Uber for driverless transportation services.

The Ford Motor Company has announced that it is working on a driverless car, a car that will have no steering wheel, no brakes, and with seats "in the round." General Motors has invested $500,000,000 in Lyft, a competitor of Uber. It is only a question of time before the General Motors – Lyft driverless car is available.

The objective is to eliminate the need for anyone to drive. The effects of this will be not only to eliminate a burden within each home or business, but more importantly to develop ways to increase the flow of traffic within the existing highway network in the country, which are currently often jammed. The driverless car will undoubtedly reduce the jamming effect. As a matter of fact, we can look forward to the day of the robot controlled single person vehicle that can hook into the highway network for transportation of a significantly larger throughput since distances between vehicles can be reduced, the size of the vehicle can be reduced, and more channels for the single passenger vehicles can be created on the thoroughfares.

In any event, there is no doubt that the number of drivers and delivery clerks and delivery personnel will be significantly reduced through the use of drones, whether land vehicles or air vehicles, for the delivery of packages.

If drones can be useful in the delivery of packages, then they can certainly be useful in the delivery of mail. They could also be used in robotic form for the sorting of mail. There are machines now that, to a large extent, sort the mail. The concept of robots sorting down to the delivery route is merely an extension of that principle.

Robots as Companions

The use of robots in just about any endeavor is obvious. Robots will certainly be useful in the home in cleaning, washing, clothes, dishes, etc., in cooking, in shopping, and the obvious one of driving. Not so obvious will be the use of the robot as the companion for the aged, and for the single person residing alone. Robots will be able to play games with such live-alone personnel, and will be able to provide them with many of the personal services normally associated with a nurse.

With advanced forms of artificial intelligence, these robots will be able to carry on conversations with the homebound. At the present time, SIRI, Watson, Google and Alexa provide a significant capability in conversation.

Robotic Reaction to Bad Behavior in Humans

Can robots intuitively detect illogical behavior of humans and respond accordingly? What if a robot caring for a normally placid patient who suddenly becomes combative in an altered mental status from low O_2 levels or sun downers? Can the robot be programmed to adapt? Of course it can. This is the whole idea of expert systems and artificial intelligence. However, in rare instance this would not be possible. Then sufficient safeguards must be introduced so that a human can be immediately summoned to take care of the situation where the robot does not have a solution immediately evident from its program. But this raises the question, of how would the human react? There is nothing to say that the human who is summoned would act in a human fashion. This sounds like a tautology but it is really not so. Human reaction to situations varies dramatically from person to person. The requirement would be to find a human who would react in a logical fashion, or in the common-sense fashion, to any given situation. Common sense is a measure of genius, and it is surprising how lacking common sense is in a common situation. So, the answer to this is you really don't know. You just have to hope you can cope with it by putting enough controls in it, and by having a coterie of humans available to cope with the situation. You might have to send a team in. But then again, the team is always available electronically. So maybe that's the answer.

Robots Helping with Homework

Let's take the case of robots helping with homework. Would they be giving the answer or assisting the student to arrive at an answer of their own? Well that's an interesting observation. It really comes down to a robot doing homework assistance the same way a parent would provide it. Some parents might be inclined to help the student by providing the answer. A smart parent would be inclined to help the student by forcing the student to think of the solution of the problem, by referring to the lesson plans, and by having the students looking back and forth at the lesson plan to find the answer to the problem. Under examination, that's how homework assistance should be provided. That's how robots, I am certain, will be trained by expert systems to provide homework assistance for students.

Robots in Occupations Outside the Home

Robots can also be used in manufacturing. It's rather obvious that they are eminently suitable for the repetitious jobs in a production line. They are also quite obviously suitable for record keeping, for the control of a production line, for the packaging of a finished product, if small enough to be packaged, etc.

Not so obvious is that robots can also be used in construction. And yet it is obvious that many of the repetitive jobs in construction can be performed by robots. Needless to say, the installation of various components in a home or factory is also amenable to deployment with robots. Such things as plumbing fixtures, tile laying, wallpaper, etc., are certainly amenable to robotic application. So too is painting.

While it would displace many jobs, the use of robots in mining will certainly be a major step forward in eliminating dangerous jobs. The robot will be eminently suitable for that. This of course would also apply to fishing where such endeavors are dangerous, as for example crab fishing in the arctic.

Farming is eminently suitable for the use of robots. The impact of this would be to increase production, and to drive down costs.

An obvious use for robots is law enforcement. A robot can just as easily do a search of criminal behavior in the past by looking at the facial parameters of a suspect.

In summary, there seems to be few, if any, occupations not amenable to significant labor and cost reduction through the use of robots.

The major impact will be the reduced number of jobs. Jobs as we know them will be significantly reduced. However, the number of jobs in certain areas will increase due to the use of robots. Before one supposes that this will be especially associated with the making of robots, rest assured that robots can make robots. Once the production model is set, it is certainly amenable to manufacturing of robots by robots. As a matter of fact, at this time, Smartphones can be assembled by a robot from selection of the individual assembly components or chips according to the formula of use of the Smartphone. There is only a step forward from there to the actual design of the components using artificial intelligence, expert systems, and the robotic generation of chips.

Infrastructure

The use of robots and control of various vehicles and devices electronically at a distance will become very important in the repair and creation of infrastructure. While this will certainly be important in the developed world, it will be without measure a tremendous capability for the developing world. This will be especially true for the new areas of the world that will be connected and wired for the first time when the new satellite systems are deployed within the next five years.

While bridges and roads come to mind when considering infrastructure for most of the world it is water that is the most important infrastructure element, Consider the impact of the ability to provide adequate water supplies for drinking and agricultural purposes. This is truly the highest priority item.

Or the creation of small bridges to make mountain passes tractable. Examples could go on and on. Roads through the mountains, which normally would be intractable. Testing and repair of existing structures. Not only would the risk of serious injury to humans be reduced and even eliminated, but actually, the cost would be substantially reduced since the use of robots would be much less costly than that of humans. The training factor alone would more than economically justify the use of robots than humans. In fact, without robots, infrastructure needs might not be met for generations. With robots, the needed programs would be speeded up dramatically.

The examples of the application areas could go on and on. There's no doubt that robots and electronically controlled machinery will be a significant benefit in the repair of existing infrastructure, updating existing infrastructure, or creating new infrastructure where it does not exist, anywhere in the world. Much will be done in the next ten years.

An Encounter with a Robot

Just think what it will be like talking to a robot. You can get some indication of that now by using the Amazon Alexa. You can ask it to play various music, for example the songs of Frank Sinatra or Bing Crosby, or any other singer that you may want to choose. Or actually asking for a particular song, or even the music of some particular musical comedy, or a concert. You can even have a conversation with Alexa concerning the history of various pieces of music. Alexa will be able to describe how various pieces were composed and also give a description of a particular piece that you may ask for. What that simple example amounts to is how a robot can interface with you and provide a concert or entertainment at your request together with explanations of different parts.

On the other hand, you might ask for a lecture on quantum computing, or on the new space satellites. Or any other topic. You could ask for Alexa to read a book to you. There are unlimited things you could ask for.

You could for example ask for the French version or the German version or the Chinese version of any book that you wanted to have read to you.

You might also engage in a conversation with your robot as to the cleaning cycle of the various parts of your residence. You might even engage in a discussion concerning repairs, painting, budgets for such items, etc. As already indicated, you might also have your robot connect to a physician on a call and your robot could be the intermediary providing connection of various components such as a stethoscope, blood pressure instrument, etc. that would be connected to your Smartphone providing direct communication with the physician on call.

And finally, you might ask your robot if you are single, and so inclined, to find you a date for the evening. Hence your robot would enter into a stream of other robots representing other individuals, and compare notes and arrive at a recommendation. That would leave it to you to communicate with that person just as you would in arranging any other blind date. In this case the robot would be the intermediary for the blind date, with the exception of the practice in vogue now, that you would have a chance to talk to your proposed blind date before going out.

Best of all, as mentioned, for the student, the home robot would be a direct assistance in doing homework. What a joy for parents, and what a joy for the students.

On a whole, the robot will be the constant companion. This will be a boom for the raising of young children. A robot will be constantly at attention beside the crib of new babies, reporting and acting immediately in case there is any difficulty whatsoever. One of the blessings of the robots will be a reduction if not an elimination of sudden crib death. Robots will also be the constant companion for the children as they grow, not only watching over them, but being with them at all times.

I could continue with all the advantages of using a robot. I can think of one major advantage that I wish I had during my career. It would be great to have a robot that would be handling my luggage as I traveled all over the world. Quite often in the past I would be on a different airplane each day of the week. To say the least it was wearing. Not having to go through the luggage hassle, not having to check into hotels and pack and unpack is a blessing devoutly to be encouraged and hoped for. Needless to say, I am really looking forward to having my personal robot.

CHAPTER 5
Robots in Hazardous Occupations

Robots can perform the physical rescue in hazardous situations. Rescue services are normally associated with floods, earthquakes, hurricanes, and landslides in the mountainous areas. These are normally associated with boats or helicopters. Ultimately, the helicopters could probably be flown remotely, although flying helicopters are a little bit difficult in complexity than flying a fixed wing aircraft. Even then, there's no reason why it could not be done remotely through electronic controls. In the same fashion, there's no reason why operating the hoist with a basket that is lowered in rescue situations could not be done totally with robots. It is expected that the first use of robots and electronic control of helicopters should be operational well within the next ten years, probably within five. It is an obvious extension of other activity. It must be admitted, however, that this use of robots would be somewhat complex and would require the development of an expert system to control the robot.

The use of robots in water rescue would also require the development of expert systems. This is certainly feasible. Perhaps the first situations where robots would be used in water rescue would include human operators as well as the robots. Ultimately the entire crew associated with water rescue could be robotic.

Rescue services are always perilous. Helicopter rescue in mountainous areas is particularly hazardous because of the close quarters, the altitude and the normally strong winds. There have been many instances where the rescuing helicopter itself was smashed against a mountain face and required evacuation help itself.

Rescue

Rescue at sea, especially in a hurricane or other forms of bad weather can also be very perilous. Even rescue boats with an experienced Coast Guard crew and specially equipped cutters can be subject to extremely perilous conditions.

Hence, in many rescues at sea, in the mountains or during floods associated with hurricanes and tsunamis, rescue by helicopter or by boat can be extremely perilous. Robots and radio control of the helicopters or boats is very much desired. It is expected that within the next ten years such control will be standard.

Hurricane Hunting

The US Navy initiated this activity years ago, but it has since been assigned to the Air Force. They now operate the hurricane squadron. This squadron flies missions in determining the strength and possible direction of hurricanes that threaten the continental United States along the seaboards. The Air Force also operates a squadron checking out hurricanes and cyclones in the Pacific.

The procedure followed is for the aircraft to fly directly through the eye of the hurricane. To say the least, this duty is difficult, requires expert training, and steel nerves. The airplane must be thoroughly checked for any structural damage after every flight, and before each new flight. Currently the hurricane squadrons use a modified Boeing 757. These aircraft are equipped with numerous electronic measuring stations within the aircraft that measure various aspects of the storm. All of these measuring stations are manned by specially trained crew.

There's no reason why these flights could not all be radio controlled and manned with robots.

Preparing robots for duty in this circumstance would require the development of expert systems. This would not be difficult since the requirement here is merely the operation of various forms of instrumentation. Flying in the aircraft is merely a variation of existing aircraft flight capability. The only difference here would be the requirement to be able to cope with the extreme set of vibrations, winds and flying through the hurricane from the entry point to the penetration of the eye of the hurricane when flight is relatively calm. This calm flight does not last long. The eye of the hurricane is normally between thirty and fifty miles across. This is only a few minutes of flying time. Then the hurricane is penetrated again and flying is rough until the aircraft exits the eye of the needle of the hurricane. The duration of that particular segment, as well the entry segment to the eye depends upon the strength of the hurricane and the dispersion of the winds. Since they are rotational in nature, to a large extent the complexity of the flight will depend upon the penetration direction of the aircraft. It can be assumed that such flights are usually associated with the maximum width of the hurricane, as such. The objective of the flight is to determine the strength of the hurricane, and any indication as to its increase or decrease in strength, speed of travel, and direction of travel.

In-flight security for the crew is not particularly hazardous but it can be very uncomfortable. Injuries are normally sustained by crewmembers that are foolish enough not to be solidly strapped-in as the aircraft penetrates the hurricane twice. The calm in the short flight through the eye of the needle can be unsettling.

Replacing the crewmembers with robots would be a relatively simple task. The various instruments need merely be operated. This is a straightforward expert system that can be duplicated in a very short period of time with minimal effort. Flying aircraft is both straightforward and proven. Flying any aircraft remotely has now been developed as an art. Flying it for hurricanes is not necessarily a simple task. It requires iron nerves, and the ability to react quickly to the pitch and yaw of the aircraft while in the hurricane. This would pose no difficulty in time to a trained remote control "pilot". It is currently encompassed in advanced automatic pilot systems. In fact, a pilot flying in bad weather often engages the automatic pilot.

It is hoped, especially by the crew of the hurricane squadrons, that the use of robots and drone flying will become standard in the very near future in hurricane hunting.

Deep Sea and Arctic Fishing

Some elements of deep-sea fishing are hazardous. This is especially true of crab fishing in the Arctic waters. Robots could certainly be used for this activity. Control of the boats could also be handled by robots and with electronic controls, much like the flying of aircraft such as drones from a distance.

Control of boats and fishing would also be a prelude to the control of boats in rescue and hazardous situations.

Such capability almost exists now. There are simulations of various types of ships that can operate in various parts of the world. Simulation capability systems can give the operator a sense of navigating an 80,000-ton cruise ship or a 40-foot speedboat through different weather conditions imposed upon the operator by an outside referee. These programs could certainly be modified so that a simulated deck and control room would actually control the ship itself. With this significant head start, it is expected that remote control of ships will be standard well within the next five years. Early adoption is expected in hazardous situations such as crab fishing in Arctic waters and in rescue in difficult sea conditions.

Fire Fighting

Probably the first instance of remote control of boats in hazardous circumstance will be in fires. Remotely controlled boats manned with robots would certainly be useful adjuncts in fighting fires that are extremely hazardous. Proximity to the fire would no longer be a serious hazard for fire fighters who would be robots. The first use of robots for such activity will come within five years if not sooner.

Fighting fires on cruise ships has always been a difficult and serious problem. Small boats equipped as fire fighters are not always available. To a large extent, ships are on their own. This is particularly true of cruise ships. These are invariably large and bulky ships, with large numbers of passengers. Modern cruise ships go anywhere in the world, including the most treacherous waters such as Cape Horn. Such ships normally carry between 1,000 to 5,000 passengers and a crew of 1,000 to 3,000. The crews of such vessels are trained more for service than for safety. It would be a major step forward in safety for cadres of robots to be deployed on all ships, not just cruise ships, dedicated to constant surveillance of hazards, most especially fire patrols to extinguish any fire. Furthermore, firefighting robots could be deployed on all ships and cruise ships in particular. This will probably come about well within the ten-year cycle of projections outlined in this book. Results of the use of robots in firefighting aboard ships should be a significant reduction of fires and loss of life. It is expected that quite rapidly after that, robots would be available for fighting fires and other hazards on cruise ships. In fact, before the next ten years is over, it is expected that the use of robots will be standard throughout the maritime industry.

Oil Rigs

Oil rigs constitute hazardous duty. Just as sailing which consists of hours and hours of boring activity separated by a few microseconds of sheer terror, just so is duty on offshore oil rigs fraught with bursts of extremely hazardous duty. At least robots now are used for underwater exploration of problems with offshore oil rigs. This task was previously executed by human divers. The use of robots should be extended beyond oil rigs to any activity that is hazardous. Human crew, if necessary, can be available in the event of a malfunction of the robots. Small dangerous situations are everyday occurrences on offshore oil rigs. Periodically danger can be extreme when there is either an explosion or some kind of catastrophic blowout of the oil wall. There is also the hazard of extreme weather, although much of that is more psychological than physical.

Helicopter landings on the drilling rig platform are normally somewhat difficult because of the small dimension of the landing area, surrounded as it is with structural outcroppings. In bad weather, it becomes even more hazardous. This would certainly be an area where robot manning and automatic electronic control of the helicopter would significantly improve the safety of the crew members. This is another target area for rapid implementation in the next ten years.

There are many boats that are used to equip the off-shore oil rig platforms. Such travel is hazardous in extreme weather as well, or in the event of explosions or fire on the platform. These too would be targets for electronic control and robotic manning.

Robotic manning of the entire oil rig is also a program that would certainly have merit and priority. It can be foreseen that within the next ten years, offshore rigs to a large extent, if not totally, would be manned by robots, supplied by boats and helicopters electronically controlled from a distance.

Infrastructure

The use of robots and control of various vehicles and devices electronically at a distance will become very important in the repair and creation of infrastructure. While this will certainly be important in the developed world, it will be without measure a tremendous capability for the developing world, especially the new areas of the world that will be covered and wired for the first time when the new satellite systems are deployed within the next five years.

Consider the impact of the ability to provide adequate water supplies for drinking and agricultural purposes. Or the creation of small bridges to make mountain passes tractable. Examples could go on and on. Roads through the mountains, which normally would be intractable. Testing and repair of existing structures. Not only would the risk of serious injury to humans be reduced and even eliminated, but actually the cost would be substantially reduced since the use of robots would be much less than the cost of humans. The training factor alone would more than economically justify the use of robots than humans.

The examples of the application areas could go on and on. There's no doubt that robots and electronically controlled machinery will be a significant benefit in the repair of existing infrastructure, updating existing infrastructure, or creating new infrastructure where it does not exist, anywhere in the world.

Mining

Mining is a hazardous occupation. It is also quite injurious to health, even in its normal functioning. Miners may encounter gas, explosions, mudslides, and other forms of destruction in the event of an earthquake, or some unforeseen malfunction or accident.

Mining normally proceeds with attacking a vein of ore or coal at the end of a shaft. Shafts are dug vertically and then horizontally when veins are encountered at various depths. At the face of the vein, miners either extract the ore or coal, whatever it might be, with picks, or by drilling holes and inserting explosive charges which blow the face. The ore is then loaded on railroad cars, which then form a "train" that proceeds to the shaft for transport of the ore to the surface. Such transport can normally consist of dumping the cars onto conveyor belts that then moves the ore to the surface. In rare instances the cars are decoupled and sent individually to the surface.

To say the least the entire process is hazardous. If there is no accident, the dust and dirt are certainly a hazard to the health of the miners.

Various forms of robots have been used in mining for some time. During the next ten years, this will intensify and it is to be hoped that all hazardous activity in mining will be relegated to robots. It is to be hoped that ultimately all underground activity will be robotic. While this would result in the elimination of mining jobs, quite frankly there are some jobs that should be eliminated. I have walked through a mine as an observer and let me tell the reader that it is a hazard just to walk through a mine. Just imagine walking through the earth at a depth of 3,000 feet, along the mineshaft with water dripping off the walls, with railroad tracks, narrow gauge of course, with illumination mainly coming from the light in your helmet.

I may also add that going up and down the "elevator" is also extremely unpleasant.

I will repeat, that it is hoped that all underground mining jobs will be eliminated as quickly as possible.

Hazardous Construction

Construction of high rise buildings with a steel skeletal structure is always dangerous. At higher elevations, the winds are sometimes quite treacherous and strong. If the steel is not totally bolted, there is a risk that it can collapse. I have personally seen a thirty-story building where the steel structure was not adequately bolted down that collapsed in a high wind. Thankfully this occurred when all the workers were off the structure.

The building of bridges quite often is extremely hazardous. I remember standing on the Koffer Dam in the middle of the Mackenzie River in British Columbia. The river current was extreme. It had been necessary to go down something like 200 feet to ensure that the foundations were properly and adequately rooted. This is the Port Mann Bridge in British Columbia, just north of Vancouver.

The reader may wonder how I come to have these experiences with my aerospace training. It had to do with the early days of my partnership with John Mauchly, the co-inventor of ENIAC, the first computer. Members of our staff had developed the Critical Path Method, which the company was now extending and applying in many different types of application areas. One of the most interesting had to do with complex construction. This carried a high risk, as well as a high profit. Very few companies had the expertise to take on such jobs. Just as a little side light on this, the construction shed for the Bank of Commerce building in Montreal, forty-four stories high, was over the foundations with a cantilever approach. Every time there was a blast in the excavation process, the whole structure would shake. After a while we got used to the shaking, or so we said!

One of my favorite places to dine is Altitude 57 of Place Ville Marie in Montreal. It is especially delightful on a cold, blustery night in a blizzard. You are in the middle of the storm but you are warm, comfortably enjoying good food and a great glass of wine.

Why is this so relevant to Hazardous Construction? Because I stood on that steel work looking into what would be the restaurant. It wasn't snowing, but I still remember the whistle of the wind.

I can strongly endorse the use of robots for such construction - bridges, high rise steel structures, cantilevers, or any kind of construction that carries an element of risk and danger. This danger can come from errors, high winds, and necessary procedures that have a certain amount of risk. Believe me, this is an understatement.

Farming

During the next ten years farming will undergo a dramatic change. Robots will take on a greater and greater part of farming. The various combinations of the devices that will be available such as driverless automobiles, self-learning robots, and electronic controls at a distance will be melded into an approach to farming that will result in significantly increased food production, and at a decreased cost.

One way or another, farming consists of the following:

1. Plowing the fields to create the furrow to accept seeds, fertilizer would be added as may be required for the produce targeted for that particular area of ground.
2. Planting the seeds along the furrows in the appropriate distributing pattern.
3. Covering the furrows if this is required.
4. Watering the fields to the extent required to augment rainfall.
5. Cultivating the produce so that weeds do not interfere. This may or may not include the use of pesticides.
6. When the produce has grown to where it can be harvest, then the harvest procedure occurs, and then the ground is repaired for the next cycle.
7. Depending upon the number of crops of the particular produce selected for this ground

area, in concert with the weather patterns for this location, then the ground to follow in the next planting season can commence if, possible, the next cycle of planting will begin immediately

And so, the farming cycle continues. It is somewhat mechanical, depending upon variable conditions such as the nature of the ground, the requirement of the ground for various fertilizers, close attention to the produce as it grows in order to maximize growth, and minimize loss due to pests, weeds, blights, etc.

All of this is mechanical and predictable but requiring control of all aspects. This can certainly be provided. If pilots thousands of miles away can control a drone, and even discharge its ammunition, then there is no doubt that farming can be controlled electronically whether it is robots or people in the actual farming process.

High-Rise Farming

There is no reason why farming cannot be accomplished in multi-story buildings in the city or on the farms. Indeed, such buildings could very well constitute the farm of the future. They can be almost totally robot controlled and worked. They would be the equivalent of multi-story greenhouses, with sunlight coming in from the side. Alternately, the buildings could be stepped. Sunlight could also be provided in a controlled manner via lamps.

Such buildings could produce more than one crop per year. Furthermore, they would have the advantage of immediacy, and proximity to markets.

The end-result of this approach will be significantly increased production, at a much lower cost. There is no doubt that the food supply can be ahead of the requirement; in other words, food will always be produced at a surplus in all parts of the world. There is nothing to impede establishing production goals that the food supply will always exceed the requirement.

Food and Restaurant Delivery

Food will be delivered by drones, and driverless vehicles. Normally this would be fresh food ordered from the food supermarkets. However, restaurants could also begin delivery of meals with driverless vehicles. In fact, a combination of robots and driverless vehicles would make it possible to have mobile kitchens circulating in high delivery areas with food preparation conducted on the vehicle. The pizza delivery systems of today are a forerunner of such mobile preparation and delivery of meals beyond just a pizza.

Riot Control

There's no doubt that robots would be of extreme assistance in riot control. They could be very effective in maintaining order, and there would be little risk of personal injury since they are robots, and not humans. Perhaps a hidden element in all this would be the reaction of the rioters to being prodded and controlled by robots. My personal opinion is that there would be a certain amount of humor associated with this. The anger normally directed towards the police would have no impact of inanimate robots. Knowing this, the rioters might just laugh instead of increasing their resistance. It is an interesting concept and will certainly be tried in riots for public disturbances in the very near future. In fact, it may already be tried through the publication cycle of this book. As of the writing, there have been no riots controlled by robots to date.

Combat

It goes without saying that the use of robots in combat would certainly reduce causalities amongst human soldiers. In addition, the robots could be used in extremely hazardous situations, and not only would causalities be zero, but because casualties amongst robots would not be considered something that would be a show stopper, objectives could be achieved that might otherwise not be possible. I keep thinking of the hazardous aspect of erecting a bridge in a military situation under very heavy enemy fire. This would be a perfect use of robots. Another would be a frontal assault, or assault upon a hill or high ground. Robots will take numerous hits before they will be disabled. Unlike a human who might be totally disabled, or even killed with a single shot.

Cyber Warfare

This type of warfare is expected to become the dominant means whereby enemies in contention would seek to achieve mastery.

Cyber warfare would consist of many different approaches, all of which are centered about achieving control over the enemy's data systems. In particular, by penetrating the control of such things as electric grid, water supply, fuel storage depots, and financial institutions, it would be possible to paralyze the ability of the enemy to continue to resist. The advantage of this approach is manifold. Most important is that following surrender; the victor can remove any impingements to the proper operation of these assets, all of which would now become available to the victor. In other words, the victory would come without the destruction of the enemy's assets, and more importantly with the victor able to use the assets for their own benefit.

There is no doubt that cyber warfare, with all of its benefits, will be the preferred means of incapacitating an enemy in future "combat" operations. The scope of this subject is so great, that other than this short description, it is far beyond the ability to consider the subject within limitations of a single book, much less a part of this book. Volumes and volumes can be written in terms of different aspects of cyber warfare.

The use of robots in combat would also change not only the nature of combat, but would heavily influence the decision as to whether to engage in combat at all. It is believed that combat would be materially reduced through the use of robots. Another aspect of combat is that it is firmly believed that much combat in the future will consist of cyber warfare. Heavily destructive combat will be avoided since history has proven that such activity becomes expensive to even the victor. There seems to be no real point in destructive combat.

Cyber warfare can be much more affective without the destruction of an enemy's assets. They would be disabled, and the enemy would be unable to proceed, but assets would not be destroyed.

It is expected that cyber warfare will be the predominant form of warfare in the future.

CHAPTER 6
Artificial Intelligence

We may consider and define artificial intelligence as the theory, development, and the actual computer systems able to perform tasks that normally require human intelligence, such as visual perception, speech recognition, decision making, and translation between languages. Artificial intelligence can also include a learning capability. This learning capability is associated with feedback as to which path is preferable. At all times, any system containing elements of artificial intelligence can accept input or verbal command whether by voice, keyboard, or a toggle stick. In a sense, artificial intelligence may be likened to the operational aspects with a human operator.

"AI" is really an expert system. There is no such thing as intelligence artificially created. Intelligence as such to discern possibilities, intuitive leaps, and the ability to judge right from wrong, or what is best cannot be artificially created. Intelligence is an innate feature of the human intellect. It can be copied but never created. No machine or program or system, no matter what it is called, will ever be able to conceive of the music of Puccini, the poetry of Shakespeare or the inventions of Edison. However, the expert systems we have and can create that function under the guise of artificial intelligence are still quite remarkable. The first form of artificial intelligence commonly in use is the automatic pilot in aircraft and in ships. In aircraft, for example, all of the control elements of the aircraft such as the aileron, rudder, and throttle are connected within a servo mechanism that is set with regard to a direction and speed requirement. This is in its ultra-simplistic description. Be assured, the automatic pilot in a modern transport aircraft is the epitome of servo mechanism logic. It can easily be replaced by a digital command structure. And it has been. The modern automatic pilot system in an aircraft can actually land the airplane, probably in a smoother fashion than the average human pilot. There are of course the exceptional pilots who can consummate an aircraft landing as smooth as the touching down of a feather. As a personal note, I have had both extremes. I've been touched down as smooth as a feather and I've been slammed into the ground, and gratified that the plane did not come apart as we landed.

Actually, the automatic pilot is an expert system. Expert systems can be created for just about anything. The procedure is to mirror the entire process, as a human expert would do it. In a sense this is a simulation of reality.

Many expert systems have been created over the years. The most complex of these has been the Link Trainer used in aviation to check out pilots for various types of aircraft. A similar simulator has been created for ships. I have personally piloted an 80,000-ton cruise ship through New York Harbor in the midst of a hurricane thrown at me by the instructor. I have also taken a speed boat through London. To say the least, it was realistic!

Expert systems are a form of artificial intelligence. Creating an expert system should be the first step in creating an AI system.

Simulations are a form of expert systems.

What is the history of artificial intelligence? Well it dates back to the 1950s. I personally delivered the first, to my knowledge, university course in artificial intelligence in 1959 at the University of Waterloo, in Waterloo, Ontario, Canada. I also had the privilege of working with Dr. Harry Markowitz in the early 1960s. Harry was the inventor of SimScript, the first language attempt at a programming system to create simulations or computer models of operating entities. Harry went on to win the Nobel Prize in Economic Sciences in 1990.

So how do you go about creating an artificial intelligence system? And beyond the definition what is it really? Let's do the latter first.

Artificial intelligence consists of trying many different paths and picking one that works. Then recording which worked. For example, instead of using GPS to pick your way across a field in a robotic car, artificial intelligence would remember all aspects of the direction followed by the vehicle. This would all be visual. So, visual phenomena would be used as milestones or guidance points along the way. You could say that this visualization would vary according to the illumination. In other words, since a path is picked in daylight would the robot follow the same path in darkness? Hard to say. What you have to do is have the robot pick its way in the dark and record the differences according to light intensity. This capability has been used in attempts to develop robots for arctic use. Everything is the same in the arctic except that there are some minor visual guides for the path to be followed. Conceivably it would be possible to develop a visual path to climb Mount Everest or any other mountain for that matter. This could be recorded and would create an artificial intelligence or AI route program. A simpler way, of course, would be to use GPS and establish the optimum path. Perhaps the best way of doing this would be to have a GPS recorder as someone manually climbed Mount Everest.

Why use robots? Obviously for supplies and for emergency evacuation in the event of an accident. To say the least there are many accidents among the hordes of people who attempt Mount Everest each year. No season goes by without a number of fatalities. Using robots would certainly reduce this number.

Robots could also be used under water. It is current practice now for Navy Seals to use motorized sleds to propel them forward under water. Equipped with artificial intelligence these sleds could follow an optimum path for the Seals.

Let me explain artificial intelligence with another example that I used years ago, in my course in 1959. Nothing has changed. Consider the requirement to find some means of covering the hand to avoid being burned. Various materials would be used to construct a glove. One such case would be a very heavy rubber glove. This would protect the hand but ultimately the rubber would be consumed if it were placed directly into the flames. If it were only a case of heat, then a certain amount of protection would still be achieved. Consider the strongest material that we know – steel. A glove made of steel would protect the hand for a few seconds before the glove becomes so hot it would eventually burn the hand. As a result, steel was certainly not a commodity to use in protective clothing for fire. And on and on, materials can be used until asbestos is applied. And it turns out to be the best of the materials in protecting the hand. By the way, I am just using this as an example. In real life, you shouldn't be so dumb as to be using asbestos products. You can create a material and see if the material creates a barrier between heat and hand. So, this would be the transfer of heat through the material. That's how it would really be done.

Consider the development of an AI system for surgery. The DaVinci surgical simulator is such a system. It is really a robot manipulated by a surgeon. The robot consists of various surgical instruments connected to robotic "arms" that function through a limited number of small holes into the patient's body. This eliminates the need for extensive openings to reach the surgical area. The end result is much more rapid healing.

Current research is directed to producing a surgical robot using only one small opening. This will further decrease the healing time significantly. The surgical robot is not an expert system since a surgeon is manipulating the robotic arms and instruments. When such manipulation is directed automatically without a surgeon present, then that is an expert system. An example of an "expert system" is the exoskeleton that permits a paraplegic to function in such endeavors as a marathon race.

So, artificial intelligence is trial and error until the optimum is found. What is important about the application of artificial intelligence rules with a computer or robotic device is that repetitive trials can be achieved quickly and with virtually minimal cost. There are many treatises on artificial intelligence.

An expert system is just exactly what its name implies. It is a system that performs the way an expert would perform. For example, we've already considered the DaVinci surgical assistant. This isn't quite an expert system since it is manipulated by the surgeon. As expert system, on the other hand, is something that would operate entirely on its own. Perhaps the most common expert system is the automatic pilot that flies an airplane as we discussed previously.

The automatic pilot was created to follow the requirements of flying an airplane at a predetermined altitude, at a speed and in a direction desired. Currently such instructions can include adherence to following a radio beam signal so that changes need not be made periodically. The modern variation can be set up to follow a course, or to follow a radio beam that varies in direction so that the aircraft would change its direction according to the signal received. The modern aircraft could also land the plane by following the glide path and reducing the air speed and manipulating the flaps all automatically. This is just about the ultimate expert system. The automatic pilot flies the plane the way an expert pilot would fly it.

The automatic pilot is an expert system. It mirrors an ability that we already know how to exercise. The automatic pilot has no innovative capability. It is only the human mind that can intuitively jump to a conclusion that has no basis in deductive or inductive reasoning. This is intuition. This is inventiveness. This is innovation. It is absolutely essential that the reader understand artificial intelligence and the expert systems that are an inevitable part of this concept and system; they have no intuitive, or innovative capability.

For example, Mozart is reputed to have said repeatedly that he composed no music but merely wrote down the tunes that were circulating in his head. Einstein has said that he invented the theory of relativity by conceiving how himself, as a light ray, would travel through space and merely record what he concluded from his observations on this trip. I can't conceive of any way of creating equations to do that.

This observation is rather important because I will allude to it in detail when in a chapter when I describe my invention of the Smartphone.

There are many forms of expert systems. War games as practiced by the military are certainly a variation of expert systems. Many of the video games that are available today are expert systems in many ways. There are numerous studies that indicate that those who are adept at playing video games actually improve their various skills associated with decision making, manipulation of instruments, the movement of controls, and on and on. The practice generated by the video game creates in the person playing the game a greater capability in performing various functions. In that sense, expert systems can be considered to form some element of an educational program.

The Link trainer has been in common use for decades. It is a simulated cockpit and a pilot sits within it and flies the aircraft. He is presented with various views of the outside world, and this can be manipulated or changed by an external controller who can simulate storms, turbulence, faulty aircraft conditions and the like. In that fashion the Link trainer can become a total simulation of the real thing and can be a checkout process for a pilot in his or her ability to handle unforeseen circumstances.

As indicated elsewhere, such simulations have been created for oceangoing and sailing vessels. These can be used for training or for checking out the capability of the person operating the simulation. This would be a very simple way and relatively inexpensive way to determine the ability of a captain to handle the ship they are responsible for.

Expert systems with regard to corporate management and decision making have also been created and used for many years. I remember playing a management game in 1957. At that time, we used a small machine that swept the market at that time, the IBM 650. This was a computer, but it utilized punch cards as input and output. The management games of today aren't much different from what they were as early as the 50's.

In such games, a number of competing pseudo-companies are organized. These companies then compete with each other setting prices, ad, and sales campaigns, and compete with the other companies for the overall market that is available. An external controller can manipulate the total market with a recession, a boom, a catastrophe, etc., and the teams must react. The results of each company – or team – will vary according to how the team reacts with changes to price and other variables. There is always a winner – a team with a greater profit than all the others. Such games are another form of an expert system.

Expert systems can be considered to be the cornerstone of artificial intelligence, but it must be remembered that artificial intelligence is much more than an expert system. For example, the exoskeleton, which is a form of robot, which controls the musculoskeletal frame of the human body, is very useful for paraplegics, or for those who have had a stroke. The exoskeleton was developed by tracing the reflexes and the movement of a body and then using this as a controlled mechanism to force the muscles of a paraplegic, which do not function to perform as if they did function. In that sense, it may be said that the exoskeleton is an expert system of the motion of the human body.

As this book is being written, a 30-year-old pregnant paraplegic has finished a marathon race equipped with an exoskeleton. From now on, paraplegics will be able to walk, race, even dance. The next step will be to integrate the exoskeleton with a care protocol.

Voice Input

Voice entry systems, as they are currently, are an operational form of artificial intelligence. These programs accept the human voice and translate it either as operational commands or into words. It would also be possible to have the voice input translate into another language as it is producing the words as in a direct transcription.

There are many systems available today such as SIRI, Alexa and Cortana. The company Nuance also produces a series of language converters called Dragon. These initial voice entry systems can also serve now as commands. Hence the combination of a robot, an expert system, and a voice entry system can provide direct voice control of some situation or robot. The ultimate robot, then, would consist of a self-directed learning program embedded in the robot and which can be manipulated by voice command over a distance.

How can this be put into use? Consider farming. Three of the most tedious, time consuming aspects of farming are plowing, seeding and harvesting. Certainly, the tractor can be controlled in the same way the car can be controlled. The driverless car is controlled by following marks on the highway and GPS coordinate directions and through the use of radar to prevent collision with other vehicles. In the same fashion, a tractor can be directed to plow straight furrow or alternately, through radio wireless control a tractor can be controlled even from the kitchen table. Its vision, of course, would be a video camera.

Seeding and harvesting could also be affected through the development of simple seeding machines, or simple harvesting machines. The Eli cotton gin is the first example of a harvesting machine. But in today's environment they can be considered totally robotic, operating on their own or under the control of an external operator. That operator might even be a robot with AI. Production in a farm would obviously increase with such devices, and the cost would be reduced, even unskilled undocumented labor costs more than a robot. Furthermore, the robot can operate at all hours whereas a human operator must break for fatigue, meals, rest periods, etc.

The promise of artificial intelligence is without limit. In the medical field diagnosis is quite amenable to the development of artificial intelligence systems. In that sense, they would be considered a form of expert systems. I remember working with Dr. Rudolph Kessler back in the 1980's demonstrating diagnosis at a distance. This has now been significantly modified through virtual reality in the form of a glove that can be used to palpate the human patient. In a diagnostic session, the patient would sit in front of a video terminal and would either attach blood pressure cuffs, temperature probes, or use an electronic glove to palpate parts of the body, a stethoscope, etc. There is no limit to the amount of diagnostic capability that can be exercised at a distance electronically!

Nano bodies are small, down to the size of molecules. They are smaller than the tip of a pencil or ball point pen. Yet they can be made as cubes with a gate that can be opened on command. Hence, they can be filled with a minimal amount of medication that can be delivered to the point of need, i.e., for example, delivery of chemotherapy right to the tumor and not throughout the body system. The result is to minimize the deleterious effect of chemotherapy on the body while maximizing the effect on the tumor.

The ability to control the flow or to measure the flow of Nano bodies in the bloodstream can also be very useful in diagnosis. For example, pressure and velocity is a hydraulic constant, especially in blood flows within the body. Hence as velocity changes in the coronary artery system then there is an occlusion at that point. This can be used as an alternative method of finding coronary artery occlusions without catheterization.

Finally, with the introduction of Nano bodies that can circulate in the blood stream of the body, changes in the pressure distribution of the flow can be diagrammed.

Nano bodies, it must be remembered, are the size of the pencil tip, and can be excreted by the kidneys. One other attribute of Nano bodies that is important is that they are ingested at a greater rate by abnormal cancerous tissue than by normal tissue. That means that Nano bodies can be accumulated within a tumor and by using visualization techniques that can display the Nano body, then the tumor can be outlined anywhere in the body.

Nano bodies can also be manufactured as cubes with a gate. In other words, this would be a six-sided figure with a gate that would open and allow the contents of the Nano body to be deposited at that point. With that capability, Nano bodies can be circulated within the bloodstream until a kidney absorbs them, and then at a selected time the gates would be opened and the medication would be deposited within the tumor.

Since the pressure distribution within the body and bloodstream can be determined without intervention, then the occlusion points of coronary arteries can be identified without coronary catheterization, which is not only a very effective technique but also a very dangerous technique.

Nano bodies can also be woven into the cloth of garments. In that fashion, you can wear your computer and even wear your interference so that the signal of your computer could not be read from a distance or from the outside. You could even carry encryption keys in Nano body cloth that would create a unique safety capability of your data.

Jobs

All of these capabilities become important when considering the impact of technology on employment. Many jobs will be eliminated, and most jobs will be simplified and speeded up significantly. Just consider the diagnosis at a distance can be the determination factor of an examination for a specific disease. Once diagnosed by the system, or by the robots, or by the artificial intelligence, or by the virtual reality system, the medical practitioner can then determine the appropriate treatment. That treatment can then be supplied by a human or by a robot.

Is this a threat to every profession? Yes, it is. On the other hand, it is a solution to a significant problem. What is that problem? Currently we have less than half the population of the world wired, we have less than 25% of the population of the world trained in some fashion. We will soon, within a decade, have the entire world wired. That means that everyone in the world anywhere in the world will be capable of connection via the internet and cell phone. This will open-up vast and tremendous new requirements for educational programs, food, services, healthcare, etc. Awareness via wired networks and Smartphones will create demand for services in much of the world where such services do not yet exist. That is the ultimate challenge that will develop from the technology tsunami of the next decade.

Virtual reality and 3D printing, augmented reality, and further technical advances in energy and basic human health will create a new world for all, but especially for those newly connected. While millions of jobs will be eliminated, millions of new jobs will be needed.

Consider the numbers. How many jobs will be created merely to meet the educational needs of four billion people! Even with extensive use of robots, this will create millions of new jobs. Add to that the new demands for healthcare, infrastructure repair or creation, plus all the other new demands by four billion people, and suddenly the enormity of the disruption of the technology tsunami will become apparent. While your job may disappear, the new jobs needing to be filled will far exceed those lost. Critical planning will be necessary to minimize the turmoil and disruption that is bound to occur. It is certain that AI, expert systems, and the new high speed computers will be major tools in addressing these problems. The chapter that follows will present various solution scenarios.

Inventing the Smartphone

I invented the smart phone. In 1972, I organized a company called XRT. I secured contracts in various elements of the financial world. First and foremost are two contracts secured in the 1970's? The first, in 1974, with additional associated contracts going on for three additional years, concerned the development of the very first electronic funds transfer system. This led to other contracts to do this for other banks. The contract was for the Girard bank that is now part of Mellon East. In addition, we were contracted by the Girard International Bank to develop systems for international banking, including electronic funds transfer through the SWIFT network.

This work in turn led to the contract with the American Express Credit Corporation to develop an online trading system for commercial paper.

With these contracts, XRT was well on the way to becoming the dominant company in real time systems for international operations involving trading, investment, debt, and electronic funds transfer. All of these capabilities are wrapped into the XRT treasury workstation, and the XRT commercial paper trading system.

As this work proceeded, we soon developed contracts with some of the largest corporations in the world, and government agencies. By the early 90s we were moving $3 trillion a day through our systems.

In all of our systems there was continual communication as well as online computer operations. I began to wonder why there was a separation between the telephone and the computer. I had had extensive training in the design of electronic computers, and I knew significant amounts about communication capabilities and techniques. I had been in partnership with the man who invented the computer, John Mauchly, the co-inventor of ENIAC; and had been a division head in the company, directed by Sir Robert Watson Watt, the inventor of radar. My knowledge of electronics was significant.

I stewed over this and in 1994 determined how I could combine the two into one instrument. I called it the CyberFone. I applied for a design patent in 1995 and on May 19 of 1995 applied for a method patent for the CyberFone. This was granted in 1998. I built prototypes, which I demonstrated and won numerous awards as best in show. By 1999 my prototype was a modification of the T-Mobile cell phone, which could fit in the hand. I demonstrated this to many organizations. I had tentative sales of 400,000 units and was well on the way until disaster struck. We had selected a chip made by one of the largest chip manufacturers in the world to transfer data back-and-forth between the telephone and the computer. It was advertised to us in the specification sheets at 12.5 Mb per second. In reality we achieved only 2.5 Mb per second, you could have a couple coffee if you waited for the telephone to work when on the computer, and vice versa. We immediately sued the company, but by the time it was resolved three years later, our funds were exhausted. It was 2002. We had just come through the disaster of the 2000.com meltdown. Nobody was interested in providing us with funds. I decided to try and license the technology. I was repeatedly told by major companies to go ahead and sue them. My patent lawyers were one of the major firms in the country. They could not be my litigation firm because of conflicts with major clients. The same was true of every legal firm I approached.

I finally made an agreement with Doug Croxall of Marathon Patent Group, a patent accumulator company. We sued 169 companies. About 100 of them settled. The others, including some of the largest companies in the smartphone business, companies that were reaping billions of dollars in profit, cried "Uncle" to the White House. They claimed that their innovative capability was being interfered by me. I, who would have invented the smartphone, was now interfering with their innovation. How ridiculous. Their greed was sickening. And painful!

The result of these negotiations was that the Administration sponsored legislation to control lawsuits. As part of this legislation, a ruling was issued stipulating that a non-practitioner could not have the rights to any patents they might own. In other words, since I was not a telephone company, I would have no rights to exercise my patents; thus allowing the infringers freedom to infringe at will. The sole inventor has been crippled by this ruling. The billions in profit from the smartphone were not enough to satiate the greed of the infringers.

I was left with patents, and a few million dollars' debt, all of which I paid in time. Billions were reaped by the infringers. The Smartphone is now the dominant instrument throughout the world and will become even more so as the entire world becomes wired.

I'm a scientist. But I'm also a human being. John Mauchly who invented computers went through somewhat the same fate. His total return for inventing the computer was $34,000. I didn't make even that much!

I sent a copy of these remarks to Doug Croxall in an email on Nov. 6, 2016. His reply the next day was "Rocco, sounds right to me...so sad what has happened to our Country which was once so welcoming of inventors...."

Am I bitter? Not as much as I used to be. I did invent the Smartphone. As my legacy, it is much more important than whether or not I am one of the technology billionaires. Although, when I carry my bag into a hot commercial airport, I wonder what it would've been like to have my own jet.

CHAPTER 7
Healthcare
Emergency Care

During the past few years, there has been a rollout of Emergency Care Centers throughout the United States. These centers are providing care similar to that dispensed years ago, by the family doctor, and more recently by the emergency rooms of hospitals. The objective of these Emergency Care Centers is twofold; first, of course, is to provide the necessary service to the patient, and second to alleviate the increasing load in the emergency rooms of hospitals. To a large extent, in many urban centers, these emergency rooms were unbelievably crowded, in many cases providing shelter as well as medical care. It was sometimes a quandary as to whether the care was more shelter than medical. This created long waiting lines that interfered with the level of care for those seriously ill.

The Emergency Care Centers were not only successful in achieving their objectives of providing both healthcare of an immediate nature, and reducing the load at the Emergency Rooms of the hospitals, but also became a financial success with investors and Wall Street.

During the next ten years, there will be a dramatic shift as these Emergency Care Centers in turn will be replaced and even made superfluous by electronic diagnosis provided in the home by teams of doctors working together at electronic centers.

These centers will be in direct communication, via the Internet or normal telephony, with the individual patients in their own homes or wherever they may be. It is obvious that in an emergency, such as an accident, medical service can be provided on the spot, and immediately, through direct connection. Such direct connection is available, even today, but immediate medical services are not always available. In case of severe trauma, helicopter service is usually provided from the accident scene to the designated trauma center for that accident location. In a sense this approach will be the new emergency rooms or Emergency Care Centers operating with much more efficiency. Augmented with robots equipped with instrumentation, diagnosis and treatment can proceed with the patient remaining at home, or at the accident scene. In the case of an accident, use of the helicopter is always an option. The impact of this will be to reduce the cost, and dramatically improve the general health and welfare of the population as a whole.

The Driverless Car Impact

Furthermore, with self-driving vehicles, or at least with all cars equipped with automatic braking to avoid collisions, the number of car accidents is expected to diminish significantly. Ultimately, it is expected that they will come close to zero. The impact of this will be significant in reducing the number of deaths, the number of serious injuries, and, for each driver, the cost of car insurance.

The Newly "Wired"

Direct connection, and the use of robots, will be most significant for the four billion people newly connected during the next ten years. The explosion in demand for health care services would be impossible to meet without this approach. Just think about it for a moment. How else can the demand be met with four billion more people added to the potential patient population?

The Affordable Care Act

In the United States at this time, the impact of the Affordable Care Act has been to significantly increase the premiums for those continuing to be covered, while reducing services available to all, even though the number of those covered has been significantly augmented by those normally incapable of paying a premium necessary, or who have prior conditions not normally covered prior to this act. With a mandate or desire to cover all citizens of the United States to the largest extent possible, the Affordable Care Act was passed with the hope of providing such services, at a reduced cost for all. The designers of this legislation believe that electronic record keeping would provide the economy of scale that would make increased service at a lower cost possible. That was the objective. It couldn't be stated well than that. In considering the use of robots, electronic augmented reality, and virtual reality in healthcare to provide service anywhere to anyone in real time, or staying well within the budget. This concept of using robots, AR and VR can certainly meet the requirement of the framers of this legislation; and even beat it!

Hospital Cost Reduction

The basic key to reducing the cost of healthcare is to significantly reduce the cost of handling patients in hospitals. With robots in the hospital and at homes, coupled with online physician centers, better, more widespread, and more immediate healthcare can be provided, probably at a much lower cost than currently. This could very well include significant reduction in hospital confinements, and quite probably in the use of hospital rooms. In fact, this could very well result in reduced demand for hospital rooms, and a searching examination of the use of hospitals.

Consider for a moment how hospitals came to be. Hospitals historically came into being for the treatment of the sick. They were designed as isolation locations to which the dying and contagious sick could be taken to segregate them from the general population in order not to spread the disease they carried. In addition, hospitals were created as a convenience for physicians to make best use of their time in traveling from patient to patient. With all the patients within one building, the travel time would be reduced. In addition, hospitals were used as training locations for the students, interns, and residents who would attend physicians as they made their rounds, discussing with these students each case as they came to it. Hospitals in this way also provided a wide variety of illnesses and variations of a particular illness for examination. There is no doubt this provided an excellent Socratic educational capability, especially in teaching and University hospitals.

Finally, hospitals became satyrs for expensive high-tech equipment, especially scanners, MRI, and radiation devices, most notably for the treatment of cancer.

Most of this will be applicable to children's hospitals, with one notable exception. Children's hospitals, to a large extent, are also intended to mitigate the fear and apprehension of children separated from their parents, and in many instances, for undergoing treatment which is disfiguring, the case of cancer treatment, or painful. Every attempt is made in hospitals dedicated to children, to give them comfort, and to provide accommodation for at least one parent or close relative to be with the child, if such a person can manage the time to do so.

With the VA Hospital, on the other hand, a hospital often provides accommodation for a homeless or destitute veteran. Currently there is major difficulty with the VA under heavy criticism for the lengthy delays in providing services to veterans. This problem could be eliminated almost overnight, through the use of electronic teaming of physicians as outlined below. Robots are available today. Robots specifically designed to provide medical services could become available quickly. There is no doubt in my mind, and hopefully in that of the reader, that the current problems preventing the VA from providing adequate medical service to our veterans can be solved, and virtually overnight.

Electronic Hospitals

Electronically, this same advantage can be continued without having a massive hospital structure operating at a high cost. Except in special circumstance, all patients could be treated in their own homes. The diagnostic instruments available for connection to Smartphones, or for the use of a medically equipped robot in the home, would allow diagnosis and treatment to be handled by a single physician or team of doctors operating from a virtual hospital. Indeed, a team of doctors associated with the care of any patient need not be located in one place. Through Augmented Reality techniques, doctors in many different locations could be consulted at the same time with regard to any patient anywhere. With AR, there would be better training, better results, faster cures, and more immediate discharge, if there were a hospital confinement. Total cost will certainly be reduced per patient. It is conceivable that total cost of all patients would actually be lower. If so, this would be the equivalent of finding, immediately, significant amounts of new funding. In other words, savings in current budgets would form new funding. This new funding will be used for additional services, rapid deployment of more robots, documentation of successes with newer techniques, and increased development of electronic record-keeping, one of the original intentions of the Affordable Care Act.

Epidemic Determination

One further benefit from this approach very well may be rapid identification of an epidemic, rapid evaluation of vaccination techniques, and rapid identification of successful treatment techniques. All of this would certainly benefit a wider population of patients, and result in significantly improved healthcare for everyone.

The "New" Hospital

With the capability of electronic diagnosis and treatment at a distance, the concept of large centralized hospitals with hundreds of patients, operating at a high cost, would become a rarity rather than a common element in the healthcare in each city. It is expected that patients would improve much more rapidly being able to convalesce in familiar surroundings, especially comfortable surroundings, with no reduction in the effectiveness of the care provided. The robots in the home will certainly be able to provide much more constant attention and care than normally provided in a hospital today.

Even hospitals that continue to exist will rely heavily upon robots. The end result will be to make the physician's time used in a much more productive fashion, improving the training in the residents, but unfortunately significantly reducing the number of

117

nursing jobs. A safe estimate would that these jobs would be reduced by at least 50%. On the other hand, there would be a dramatic increase in the training requirements for the nursing jobs that would continue in such countries as the United States. But the number of nursing jobs required for the four billion people newly connected will be significant. It is conceivable that the number of such jobs will exceed the jobs lost through the adoption of electronic technique and the employment of robots.

Implementation Schedule

The most important fact that must be taken into account is the speed with which these developments will come about, even in the United States. There will certainly be some resistance to change. However, it will not be long before the significant benefits of these changes will become readily apparent and will act as a catalyst to make change more readily acceptable. Furthermore, the demand from countries that have poor healthcare services at this time will generate significant new jobs; and these jobs may not require movement or relocation to those areas since services can be provided electronically from anywhere to anywhere at any time. The most optimistic approach would be for those whose jobs are dislocated in the United States remain in their current location, while still having jobs to provide services in other countries of the world. Just as pilots located in Las Vegas, Nevada control drones flying in countries of the Middle East, in the same fashion it will be possible to provide nursing and physician services from locations in the United States and other developed countries to the newly connected countries of the world. Remember even the language barrier will be overcome with forms of artificial intelligence that will allow immediate translation. A significant number of languages have this capability at the present time, but new languages will be added very quickly as the robots themselves become learning systems as well as teaching systems. This point will be discussed in the chapter on education.

Export of Healthcare

Healthcare, including the education of healthcare professionals, could even become an export product for the United States. Income generated from such endeavors would certainly be welcome as additions to the healthcare budget. Such income would be earned under contract to international locations.

Just for a moment consider the possibilities now of providing services from a location in the United States to monks in Laisha Tibet, or to a remote village in the Belgian Congo. A physician or a group of physicians in the United States in a single day could be treating patients in the United States, deepest Africa, and the Mongolian desert, France, Antarctica, and New York. Such distribution of patients would provide no difficulty whatsoever to the system. Through electronic means, augmented with robots attending a patient, diagnosis and treatment would proceed as if the patient and the physicians were all in the same hospital settings.

An exception that can be envisioned at this time has to do with radiation. It will still be necessary to transport the patient to the point where such extensive radiation capability exists. However, it is conceivable, that radiation as a means of treating cancer will diminish in importance, and might even disappear. At the current time cancer is treated by burning it with radiation, poisoning it with chemotherapy, or cutting it out with surgery. A more dramatic way, and a more successful way, of handling cancer has emerged in the last ten years. That is the boosting of the immune system of the body so that the immune system itself destroys the cancer. Cancer, it must be remembered, is an aberration of healthy cells in a particular organ. In other words, liver cancer and bladder cancer and kidney cancer are all different diseases. In each case, it is the cellular structure of a particular organ that begins to generate bad cells that we call cancer. This is similar to weeds in a garden.

At this time, a number of cancers are being eliminated through the boosting of the immune system by injection of serums that include Interferon 2 or some other carrier for the treatment, cancer cells from the particular tumor, and some catalyst pertinent to the organ and the cancer. For example, the catalyst for lymphoma is thalidomide; the catalyst for brain cancer is a variation of the polio virus, and so on. It is expected that within the next ten years the number of these remedial medication techniques to boost the immune system will proliferate and will be used in more and more cancers. The end result will be significantly fewer surgeries, the potential elimination of all chemotherapy approaches, and the significant reduction if not elimination of radiation. The treatment of cancer will evolve from one of destruction to one of boosting the immune system of the body itself to destroy the aberrant cells constituting the cancer.

Security with Electronic Devices on Your Body

Currently there are many electronic devices that are implanted or used on the human body. Most notable of these are heart pacers and hearing aids. With both it is possible to have external control. In other words, the heart pacer can be connected to external cardiac centers for examination or for changing of its settings. The same is true of hearing aids. These can be manipulated electronically either with devices that the patient can use to change the program, or the volume level. These can also be manipulated by the audio specialists in their office. During the next ten years, the ability for the audiologists to regulate your hearing arrangement without you going to their office will also be pioneered.

The arrangement for heart pacers is currently available. This offers a tremendous advantage to the patient by not having to proceed to an office in the event of a malfunction. It offers more safety in health as well. Unfortunately, just as everything else electronic it offers a target for hackers and exploiters. Conceivably, a hacker could take over control of your heart pacer, your hearing, car, bank account, Smartphone, or the alarm system in your home. It goes on and on. Security measures become an important element of everyday life, as it will be lived in the next ten years, and thereafter. For now, it must be assumed that your heart pacer, your hearing aids, car, and home will be suitably protected with counter measures from attempts to hack or pirate or take them over.

These security measures will be considered in greater depth in another chapter of this book. For now, it can be assumed that you are protected. In the final analysis, you can always opt not have these external control capabilities by having a heart pacer that is not connected. Hence it cannot be hacked.

Cancer Treatment

There is little doubt that immunotherapy will rapidly become the treatment of choice in cancer. While there certainly have been successes, temporary if not permanent, with surgery, radiation, and chemotherapy, these are somewhat brutal and often painful approaches to generating a cure, I can attest, on firsthand authority. The pain associated with some of these treatments is at least a 10, and would be higher if higher numbers could be used. The only danger, as such, associated with immunotherapy could occur if the immune system is so stimulated that it not only knocks out the cancer, but also then proceeds to knockout the patient. In extreme cases, death can result from the immune system going berserk. One such example is the hazard associated with the use of bone marrow transplants in the treatment of pernicious anemia. However, it must be assumed, that with significantly increased application of immunotherapy techniques, the countermeasures will also be developed with regard to reaction. In any event, despite occasional setbacks, there is no doubt that immunotherapy will become the standard in cancer treatment and that surgery, radiation, and chemotherapy will diminish in application use, and conceivably could disappear entirely.

Nano Bodies

As mentioned, Nano bodies are small particles, smaller than the size of the tip of a sharp pencil. Nano bodies can be manufactured as cubes with a gate that can be opened and closed with an electronic signal. Hence, such receptacles can be filled with a particular medication, loaded antibodies, which can then be delivered to the appropriate location with the medication required.

Once a sufficient quantity of the Nano bodies is present, an electronic signal would then open the gates and the medication would be deposited at that location. Guarantees that the medication can be delivered exactly to the point of need would have a twofold major beneficial effect. First of all, the amount of medicine required would certainly be less than what is normally sent through the system in the hope that it would affect a particular point by circulation through the entire system. Second, the impact of the medication can be more readily assessed since there is complete knowledge as to how much and when medication was delivered exactly at the point of need.

Furthermore, there is one further major benefit of this approach. If the DNA of the patient were known, it would be possible to assess, actually directly, the impact of medication links to the DNA of the recipient. In that fashion, it would be possible to determine what medications should not be used according to the DNA structure of the proposed recipient. On the whole, this approach would result not only in more immediate cures, but also correlations of medications with DNA. Medicine will take on a much more personal nature as medication would be directed according to specific DNA reaction and not, as is the practice today, of using various medications according to average responses, independent of the DNA of the recipient.

The administration of medicine will also be directed more to the location of where such medication is needed, as well as the selection of medications that will have the maximum impact taking into account the DNA of each patient. Medicine is in this way much more personal and accurate.

Healthcare – An Assessment

On the whole, the field of medicine will undergo a significant improvement in the capability of curing disease, and of treating patients in their own home settings without the need of moving the patient to a hospital setting. Remember that many patients become more ill in a hospital because hospitals are accumulations of many diseases all in one place. As hospital personnel know today, hospitals have become a perfect place to spread a disease very rapidly. Allowing the patient to stay in their home, and providing both isolation and improved treatment, would significantly improve not only the health of the patient but also that of the general population as a whole.

The end result of all this is would be improved health care being provided to a much larger segment of the population at a much lower total cost.

The combined elements of constant service are vital. In the Affordable Care Act, more is added to the cost as more people come into the system. The original objective was to cover the entire population, and at a lower cost for everyone. With current practice this is an oxymoron. However, with improvement it might be possible. The original plan called for extensive use of electronic records. That may be so, but the manner presented in this chapter offers a much greater possibility of meeting the requirements of the entire population, hopefully at a lower cost. That is, to have service provided electronically through extensive application of augmented reality and virtual reality with a link between patients and a doctor or group of doctors. Patients and physicians need not be at the same physical location. That is the key to this concept. Furthermore, electronic records are certainly a viable byproduct of this approach. The end objective, it must be remembered, is not electronic records, but treatment resulting in the care of the patient. That was missed entirely in the background consideration, which led to the development of the Affordable Care Act. Providing healthcare treatments outlined in this chapter will go a long way towards making healthcare affordable, effective, and capable of taking care of the entire population at a total cost which is tolerable. However, the use of robots in this milieu will be both beneficial and cost-effective.

Meeting the Budget of the Affordable Care Act

One way or another, it is guaranteed that the cost of the Affordable Care Act will be spread among fewer and fewer patients paying full insurance, whereas the number of those receiving either free care, or cash supplements for their insurance premium will increase. As a result, those paying the full fee will decrease in number, and that full fee will increase in the percentage of the total population. Hence, those that who were paying, will pay more.

This is an unsatisfactory situation, and one that runs contrary to the promises made when the Affordable Care Act was considered for passage in the Congress. Citizen support was solicited through promises that the total cost would be reduced, and that every citizen would be allowed to keep their current insurance if they so desired. This turned out to be falsely stated. Quite often the current insurance has been totally withdrawn, as the companies in question have decided to terminate their coverage in a particular state. This was allowed in the act. This was a mistake.

Through the use of the physicians contact and care centers, operating as primary care physicians, together with the use of robots and the implementation of devices attached to smartphones for use in diagnostic procedures, it is expected that the services needed to make the Affordable Care Act cost affective will be implemented. In other words, it is expected that the physician care centers and the diagnostic capability at a distance will reduce the cost of care for each instance where it is required. Furthermore, as already outlined, it is expected that the emergency centers and emergency room usage will drop dramatically. All of these factors will lead to significant reduction in the cost of providing services with the Affordable Care Act. On the whole, it is expected that the Affordable Care Act will now become affordable. Furthermore, it is expected that the number of people that would be covered nationally will rise to close to 100% of those who might require health care services. This was the initial objective the Affordable Care Act, and it would appear that this objective might very well be met through the use of electronic care centers and the use of robots.

Veteran Administration

The VA has come under increasing criticism during the past two or three years. Waiting times for veterans has increased disgracefully. Quite often it takes months to get an appointment, and there have been many instances the veteran has died before they can be seen.

In two instances the secretary of veteran affairs has been replaced dramatically. Apparently, the problem persists.

The problem could be solved with the methodology already described. A combination of electronic call centers with teams of physicians available via electronic means, and robots dispatched or assigned to the home of the veteran would go a long way of alleviating if not eliminating this problem.

The electronic call center could serve as the primary care physician in establishing what kind of specialized care, if any, is required by the veteran. The robot assigned to the home of the veteran could provide support services augmented, if necessary, by a visiting nurse, or by physician or physician assistant.

Similarly, a diagnosis could lead to hospital confinement if that is necessary. As already outlined, to a large extent the traditional concept of hospital confinement often leads to even more difficulties than it solves. As already outlined, through the use of on call assigned robots, working full time with the patient in familiar surroundings-the home, together with immediate access to on call physician groupings should alleviate the back log of the veteran problem very quickly.

If nothing else, this two-pronged approach should lead to rather immediate diagnosis of what is required to reduce, if not eliminate, the long waiting times currently being experienced.

Implementation Schedule

Robots in the home acting as care givers, and as nurses for diagnostic purposes, together with diagnostic instruments, which can be attached to Smartphones and then connected to the electronic physician, care centers will certainly come to fruition. It will not happen overnight, it is expected that the full ten-year cycle will be required for introduction for close to full operational capability, at least in the United States. From a practical point of view and to take into account many introductions of this nature, it is expected that the implementation schedule will follow a bell curve. This shows that there will be a steep rise at about the five-year point where the implementation schedule accelerates.

Robots, Triage and Euthanasia

In healthcare, can robots, evaluating a patient and monitoring a patient, be programmed with a sense of urgency that a human care worker would experience in the event that a routine situation turns into an emergency situation? Of course. This is the whole objective of expert training. If the robots encounter a situation that it is not programmed to handle, it would defer to an electronic medical group. This group, consisting of physicians and care givers, available electronically in the event of an immediate emergency, would be able to cope with a healthcare situation that a robot would be unable to resolve from it training. This in turn would generate a learning situation and which would increase a background capability of a robot, and of all robots with the same program. In time, it is hoped that all such situations will be covered.

One great difficulty will be the probability that robots, especially in totalitarian regimes will be programmed to terminate life rather than protect it. There will be an effort to increase euthanasia amongst the aging and infirmed. Robots might very well be programmed to terminate any life that shows no indication of improvement from a situation of a vegetable existence. And yet, there are many instances of vegetables suddenly becoming totally coherent and once again full of life. Unless a robot is trained to respect life, robots will become the instrument of justifying governmental euthanasia, often motivated by a desire to reduce the budgetary cost of caring for such people. This would be tragic.

With miniaturization of computers and technologies it is possible that scanners and other health care equipment might become mobile, not attached to someone's Smartphone, in order for the patient to be able to avoid going to a medical center at all. Of course. It will not be the miniaturization so much as the cost factor. Scanners are not necessarily mobile devices because of the manner of which a scan occurs. If the same capability can be achieved without the ring of x-ray or magnetic response signals that are interpreted as changes in density, relating to the activity within the human body, then of course this might be a portable piece of equipment.

The most important aspect of this, however, is that such items of equipment might be available for patients or cultures in more remote locations throughout the world in order to improve their health care. This might be similar to a MASH unit but on the wellness and screening end instead of the surgical end. This can be one major benefit of the distribution capability with modern equipment.

The key elements for the future of healthcare are electronic means of communication, testing, and treatment. The combination of the Smartphone, roots and instant communication will provide the mechanism to radically alter and improve worldwide healthcare.

CHAPTER 8
Education and Teaching in a New Environment

In the next ten years, we will return to the traditional way of teaching, but in a very non-traditional way. It will be the equivalent of the Socratic Method where the students gather around the teacher who fires questions back and forth in order to make sure the basics are fully understood in terms of how they are applied; but it will make significant use of Augmented Reality (AR) and Virtual Reality (VR), and will be centered on the use of the Smartphone. Robots may even be part of the scenario, either as a separate device and microchip, or with the Smartphone fulfilling that function.

To some extent the modified Socratic Method is used in the seminar approach at most senior classes at the university level. Oxford in particular has made this a standard. At Oxford, the students study on their own and periodically meet with the professor in order to gain guidance. They are said to "read" for their degree. Periodically as a group associated with any particular professor they also meet in a seminar give-and-take type of forum.

In the next ten years, the basic elements of electronic education, as such, will be achieved. This will be to apply a combination of the mechanical aspects of learning, which can proceed at the individual pace of each student without slowing down or annoying the students who are more gifted than others, and can proceed at a much faster pace overall. This will be done with robots, or with software associated with portable computers, most likely the Smartphone. My first prototypes of the Smartphone, built in the period from 1995-1999 were designed for educational programs. Education was selected as the first area in which the electronic impact using the particular capabilities of the Smartphone even in the early days when prototypes were large and clunky.

Then, and most especially in the next ten years, each lesson plan will be included within the Smartphone, and will be covered by each student separately with the assistance of a robot. The robot need not be a particular device, but could also be an enhanced software program residing in the smartphone. It would be interactive with the student, utilizing the techniques of artificial intelligence and expert systems in order to ensure that the student becomes well versed in the mechanical aspects of the particular lesson. For example, at the lower grade level, this could be in the form of drills for the various arithmetic procedures. In the lower levels of the middle schools, this could be drills in various aspects of algebraic manipulation, utilizing algebraic symbols to introduce the student into abstract symbolism, and the beginning of logic. This method of manipulating abstract quantities not only covers mathematics, but also automatically introduces the student to logic, and methods of coping with logic utilizing algebraic notation. This can be considered a form of deep immersion similar to the concept of deep immersion in studying any language. The objective is to have a student immersed totally in that language for lengthy periods during the day. Similarly, the promotion into logic using algebraic notation will go a long way not only in teaching logic, but also in converting all logic exercises into a form of language, and coping with these logic exercises through deep immersion. In the upper levels of mathematics in high school and the lower levels at the university level, this deep immersion process with logic can be applied to most subjects. For example, this could be applied most especially to the mechanical aspects of calculus. It could also cover the fundamentals of chemistry, physics, and just about any other subject.

With regard to languages, it is obvious that this approach would be of significant benefit with grammar. Because of the interactive aspect with the smartphone or with a robot, speaking drills would be significantly enhanced as conversations can take place between the student and the robot, whether the robot is a particular device, or a software aspect of the lesson plan. In any event, direct conversation can become a significant part of language training, just as it is today where language training is best accomplished with full immersion. Languages are best learned through use. It is ridiculous to think that languages can be learned by having drills on irregular verb formats. In my own case, I think I learned every irregular verb format in French, before I even understood what the pluperfect subjunctive was. In fact, I can almost say who cared? The objective was to learn to speak French. Ultimately, we did, but no thanks to the drills on the irregular formats of French verbs. Consider trying to teach our young children how to speak English by having babies go through regular drills of the irregular verbs in English. How foolish. Babies learn to speak English through osmosis and being totally immersed in English.

The robot and the smartphone can provide this deep immersion capability for any language, and as a matter fact, for any subject.

This would be the equivalent of full immersion, which has been established as the best way for a language to be learned, and to be spoken. Full immersion could also be of great assistance in teaching the grammar aspects. We learn our grammar through the speaking aspect of our own vernacular languages as we absorb them as we grow. Then it is all consolidated as it is explained to us when we study grammar just exactly what we've learned by osmosis. We don't learn a language by doing irregular verbs. With regard to a foreign language, this robotic approach would not only help with the speaking aspect, but through continual use of the language, the grammar parts would become readily apparent. Vocabulary would also be built somewhat just through the approach of speaking often and at length.

And so, we have the robotic approach, as such, directed to learning the "mechanical" aspects of any particular subject, no matter what it might be.

When it comes to the upper levels, as such, of university courses in philosophy, psychology, and most aspects of medicine, the Socratic approach of an interchange between the student and the teacher would be appropriate. The teacher can be a robot or it can be an actual person. That person can be in the same room, or can appear to be in the same room through Augmented Reality. In that fashion the best teachers of a particular subject can be anywhere in the world and be instructing a class composed of students in turn can be anywhere in the world. So too can the exchange between students, who need not be sitting beside each other, but with AR and VR can appear to be so situated, no matter where they are.

Perhaps you can envision the educational process that will be initiated in the next ten years. Students will proceed at an individual pace, being members of virtual classes. They will also be instructed by a number of teachers, some of who are robots or extended computer programs operating on Smartphones. Students will proceed at their own pace until they arrive at a certain level, at which point they will become engaged in an exchange between students and teachers, or the teachers may be a number of specialists in a particular discipline, located in different parts of the world, but appearing to be in the same room as the students, who in turn may be located in many different parts of the world but have been brought together as a group, via AR and VR, for this training session.

In this fashion, it can be said that the best teachers in the world can be made available to everyone in the world, even in the remotest of parts of the world.

Since this system of introductory lesson plans will depend heavily upon the machines - smartphones or robots - and perhaps interchangeably, then even those older people who have not had the benefit of education when they were younger can now take courses that they wouldn't ordinarily be able to attend, follow, or even understand. Even the mentally or physically handicapped will no longer have restrictions on what they might be able to or entitled to follow. What a boon to everyone in the connected world, and that will soon mean everybody, capable of at least learning how to read and write. More importantly, it need not end there.

The final result of this process will be to impart a highly personal nature in the educational process. No one will be left out. Those requiring specialized assistance can have such assistance provided by specialists, who in turn would be able to provide help to a number of students, even when such students are scattered in different locations.

The concept is very simple. The entire process goes on as follows:

1. The methodical aspect of any subject is covered through a one-on-one relationship between robot and student. The robots can be program aspects of the normal lesson plan. This one-on-one relationship is heavily verbal and students proceed at their own pace.

2. At certain points in the training program designated in advance, when a group of students has arrived at a level of capability established for that datum point, then all the students who are at that level, or at least a group of them, can be brought together into a simulated class even though they are not necessarily in the same physical location. They can be instructed by a teacher, in combination with robotic assistance, or with a group of teachers, all of who may actually be in separate locations. The objective is to establish an interchange between students and instructors in order to examine the ramifications of the basic material they have covered privately. A secondary objective is to broaden the scope of understanding of the subject matter of how it can be used, to the extent that the students can achieve such an understanding. Not all students will have that capability. All students, however, should develop and have some such awareness. As these seminar type sessions continue to occur in the training program, most certainly, this awareness of the broader scope of the

material being covered should ultimately become fully understood, hopefully by all the students, but at least by the majority of them.

3. The students are encouraged to have exchanges amongst themselves before any final or mid-term examination takes place. And by the way, examinations are a combination of written and oral and are taken when the student is deemed prepared for a particular datum point. In that fashion, students can proceed largely at their own pace, but actually in depth as if they are in a physical class with physical teachers present to direct them and assist them. It is expected that each student will proceed further with each subject than would normally be the case in our current educational approach of group learning. In fact, the current move towards a Common Core is almost the surrender of maximum achievement for every student. Common Core is almost the surrender in selecting achievement levels that everyone would reach, even if the achievement level were lower than what might be achieved individually, or even by the entire group.

Stop for a moment and imagine the impact of this recommended approach. Can you have any doubt that education will proceed at a faster pace, at a deeper level for each student, and with making greater use of the time of a teacher to teach as opposed to running through mechanical aspects of rote learning? The machines can do a better job of rote learning than an impatient human.

The impact of this will be to widen the range of coverage and impact of each teacher, but more importantly to provide the capability and assistance of highly qualified teachers in the remotest parts of the world where normally such qualified teachers would not be available. In a sense this will then become a bootstrap operation. Qualified teachers in the remotest areas would be created over a period of time through this method. This cascading effect would be the most rapid and most cost-efficient means of educating the vast number of people that would be added to the population of the world requiring education and training. Just imagine the impact of adding four billion people to the population to be educated.

When incorporating more technology into education, there are subtle factors to consider. For instance, if a student does the one-on-one lesson plan and review with a robot on a daily basis, how does the student learn to do group projects or handle socialization, which is a side effect of being in a classroom? That has already been covered to some extent. It should be emphasized, at this point, that group projects must be a part of the educational program in order to ensure that socialization does occur between students. One of the great problems associated with technology is that people tend to become more tuned into their iPad or iPhone rather than to the person sitting beside them. I know of cases where people walking down the street are texting each other rather than talking. This is ridiculous. People have to learn to talk to each other and not just to text.

Anything that encourages effective socialization and cooperation certainly is to be emphasized. Hence, perhaps the answer is to develop projects that must be handled in a group fashion. One of these, of course, is to simulate the operation of a company. Even at lower levels of the educational program, management games where the students act in different capacities in different situations certainly is to be encouraged. Instead of just corporate structures and companies that compete with each other, this could also be historic situations where the students would take on different roles. For example, it could be a dramatization of the controversy between a flat earth and a round earth, and the presentation by Christopher Columbus to Ferdinand and his followers in Spain that the earth was round.

The concept of the flipped classroom that some schools and teachers are using is an important element to be considered in any such endeavors.

There's no limit to the levels to which such educational systems can be applied. And the ultimate, if you think about it, would be the equivalent Oxford approach, no matter what subject, and no matter what level. At Oxford, this is practiced at every level. And in most graduate schools in the

United States this is the way in which most doctoral programs are conducted.

The most important aspect of education in the future is not the educational process itself, but the planning process. Currently there is a significant emphasis being placed upon a common core. This concept should be immediately rejected. There is no need to have a low level of common capability if everyone can arrive at their maximum achievably level through the use of interactive, real time one-on-one, robotic-human approach to teaching. Hence the most important aspect would be to plateau levels at which examinations would be conducted and so that students would move on to other levels at their own pace. This factor that I believe would result in making education somewhat of a competitive game. The students, and the entire community, will benefit greatly from this approach in education. The benefits would be:

1. Everyone will proceed to a higher level than would normally be the case.
2. An extended teaching cadre would be available sooner than would normally be the case.
3. There will be a highly personal level of education even though students may not be in the same physical location, and teachers may also be separated by distance. And yet both students and teachers would appear and feel that they are in the same room through the use of augmented reality.

The end result would be highly enhanced and advanced educational programs at significantly lower cost than is the case today. Furthermore, the time involved in education will be significantly reduced. On a calendar basis, there would be no need to stop the educational process for three or four months during the summer season of that particular geographic location. Classes would continue year-round unless they are affected by such requirements as sowing, reaping, and other requirements.

The combination of machines and people and electronics will be advanced continually towards achieving an equivalent of the Socratic process in the 21st and later centuries. Furthermore, everyone will have the ability of being educated to their maximum capability – the young, the old, the handicapped, and even the mentally handicapped.

When it comes to teaching advanced subjects, the benefits will become evident very quickly. For example, in presenting advanced topics in aerospace engineering, it would be possible to have the leading experts in a particular subject available via AR. The same would be true of the advanced topics in medicine and healthcare, or just about any subject to be considered.

There is one very simple question to continually ask, especially in teaching the more advanced topics. The question is: who really is the teacher? If the Oxford example is used, it is the student who is the teacher and the professor who is only the catalyst in the educational process by pointing to directions to examine. In this modern approach, it is the Smartphone that will rapidly become the major support in expanding not only the scope of the educational process, but also its depth of understanding within each student. That, fundamentally, is the objective.

The Robot-Student Interface

Anything that encourages this certainly is to be emphasized. Hence, perhaps the answer is to develop projects that must be handled in a group fashion. One of these, of course, is to simulate the operation of a company. Even at lower levels of the educational program, management games where the students act in different capacities in different situations certainly is to be encouraged. Instead of just corporate structures and companies that compete with each other, this could also be historic situations where the students would take on different roles. For example, it could be a dramatization of the controversy between a flat earth and a round earth, and the presentation by Christopher Columbus to Ferdinand and his billa, rowers of Spain, that the earth was round.

There are many situations throughout history where it would be possible to have such dramatic role playing by students to encourage socialization amongst the students as a side effect of being in a classroom.

The concept of the flipped classroom that some schools and teachers are using is an important element to be considered in any such endeavors.

The Role of AI, VR, and AR

First and foremost, is it the Smartphone that will be the cornerstone of the educational environment that will exist in the next decade? The Smartphone will capable of adding appointment that will materially enhance its range of application. For example, in considering healthcare, with the individual Smartphone would be capable of having adapters inserted in its external jack location for blood pressure, temperature, stethoscope, etc. In the same fashion, for educational purposes, hardware units, such as a helmet can be added for virtual reality. Furthermore, Google Glass or the equivalent can be added to visualize the screen. Alternately, adapters can be added to project images on the wall. Hence, smart phone can act as the central coordinator for the use of virtual reality in the classroom.

I can't tell you how important this. Just imagine the impact of having students actually visit Independence Hall, via virtual reality, when studying that aspect of American history. Or, going on a tour, via virtual reality, on Old Ironsides when studying the history of the birth of the US Navy.

Via augmented reality, it would be possible to have international classes with students in different parts world, all appearing to be in the same room. Alternately, teachers from other locations would appear to be teaching the class.

Because of the Smartphone virtual reality in augmented reality will be a major capability in the classroom of the next decade.

Artificial intelligence software products would certainly be an inherent part of any educational program of the future.

These tools, together with the use of robots, will enhance the educational process significantly. Furthermore, because the cost of the Smartphone will be much less than $10 per unit, this approach will be economically feasible in the Third World, which will now be wired. In fact, without this reduced cost of the Smartphone, it would be difficult to envisage how the educational needs of the newly wired could be met. With the Smartphone, these needs cannot only be met, but they can be met without imposing an impossible economic burden.

College Costs

The cost of a college education should be materially reduced in the next decade. Why? First of all, many courses will be available electronically. Delivering courses electronically certainly reduces the total cost of operating colleges. Electronic courses do not require buildings or classrooms. Teacher time is maximized, and then faculty salaries may be significantly reduced in total in the college budget. Furthermore, programs to other colleges can be a source of income. This can be particularly important in meeting the needs of the newly wired Third World. It may not have a college infrastructure, but will certainly have the demand and the need. College courses could be exported for their use.

Common Core

Currently, in the United States, public education is in crisis mode in many inner-city areas. In order to ameliorate this situation, a program known as Common Core is being advanced as a means of coping with some of the current problems that exist. It is the premise of the author of this book that the road to improved public education lies with the introduction of the techniques outlined in this chapter, and not with an artificially derived lower level of achievement that is synonymous with the concept of Common Core. Hence, with the use of the types of educational programs outlined in this chapter, it is expected that educational achievement in the public schools, and indeed in all schools, will be markedly improved. So too will the total cost be reduced for providing educational programs in the public sector. In the private sector, achievement levels will not only go up, but costs will go down dramatically. In addition, it also appears, public and private, and primary, secondary, and college-level educational programs, costs will be lowered. The current confiscatory levels of higher education costs will no longer virtually bankrupt the middle-class family. It has been advanced that the government should assume the burden of these excessive costs in higher education. This will become unnecessary if the procedures outlined in this chapter are applied.

A Final Word

There is an ancient proverb that goes: "The more things change the more they remain the same." The tsunami technology onslaught of the next decade will introduce major changes into the educational approaches at all levels. The classroom, whether it is real or virtual, electronic, or actual, with or without cell phones, artificial intelligence, virtual reality, augmented reality, or just a point teacher, will in its final analysis be more in tune with the Socratic process of 2500 years ago. And such is progress.

CHAPTER 9
Space and the Computer

This is an optional chapter for the reader. It is somewhat technical, but rather important to understanding the space efforts of the next ten years.

During the next decade, it is quite certain that the first voyage to Mars will take place. How this will be done, and its impact upon the Earth, is a fascinating story. As a major step forward in the history of the human race, it is recommended for greater insight. While this chapter is concerned with technical matters, the technical points covered would be understandable to high school students. Hence, while reading this chapter is an option, the reader is strongly urged to do so.

It was with the advent of the computer that space flight became possible. The first attempts were the Sputnik series by Russia and their Vostok 1 carrying Yuri Gagarin as the first man in space.

Sputnik 1, was the first artificial earth satellite. It was launched by the Soviet Union on October 4, 1957. It was a 23-inch diameter polished metal sphere with the weight of 183.9 pounds. It had four external radio antennas that broadcast radio pulses. These could be heard throughout the world. The Russians used it as a propaganda weapon as well for the scientific measurements. It was visible everywhere on earth and its radio pulses were detectable. It was launched during the Geophysical

Year, and came as a complete surprise to the United States and the World.

The US was heavily involved in a bitter Cold War with the Soviet Union. In an attempt to surpass them, even after Sputnik, the US initiated a series of programs aimed at not only matching the Russians, but surpassing them. The NACA was revamped and reorganized as NASA. An energetic scientific program was initiated, supported to a large extent by budget money from the military. US efforts concentrated on attempts to launch a satellite for the Geophysical Year. This led to and was the rushed version of the first Vanguard satellite. Full of confidence, in an attempt to upstage the Russians who did everything in extreme secrecy, the launch was scheduled for broadcast on national television. The date was December 6, 1957. It ended in an explosion. The Naval Research Laboratory Vanguard test vehicle three was a small satellite. It was designed to test the launch capabilities of the three-stage rocket. At launch, the rocket rose about 4 feet off the ground, then fell back and exploded on the launch pad. This was a hurried and ambitious Navy project to test multiple stage rockets, and to upstage the Russians. It failed.

The Soviet launch of Sputnik 1 was a scientific and propaganda success. However, it had the effect of awakening the slumbering giant. Ultimately, the efforts of the US were infinitely more successful, but the successful launch of

Sputnik I certainly caused consternation in US military, scientific, and political circles.

The catch-up program had been intensified with the launch of Sputnik 2 on November 3, 1957. It had a weight of 1120 pounds, and contained a number of separate compartments. A dog, initially named Laika, a stray dog from the streets of Moscow, became the first living creature known to fly in space. The initial program was for a ten-day trip but the dislodging of some heat dissipation tiles on launch caused the internal temperature to rise to more than 40°C, or 104°F. It is known that the dog ate its food and moved around the capsule, but it survived for only two days. The scientific monitoring and measuring equipment on board gave the Russians sufficient information to attempt the launch of a human. The Vostok 1 was launched on April 12, 1961. It carried the first human in space, the Russian cosmonaut Yuri Gagarin. This was part of the Russian space program of six manned flights from 1961 to 1963. No further manned space flights were made in 1964 and 1965 by modified Rostock spacecraft. By the late 1960s, the Russian Soyuz spacecraft was developed. These are still in use today shuttling back and forth to the International Space Station.

The first American launch attempt, the Vanguard, was a dismal failure and was destroyed on the launch pad. There are various stories connected with this. One is that it was a perfect takeoff but that the visualization was wired backward so that the controller thought that the rocket was coming back to earth and initiated the destroy signal. Whether this was true or not it was a tremendous setback for the American dream of beating the Russians.

NASA was formed in that year, and launched the first American satellite. From then on NASA was the world leader in space, ultimately landing two men on the moon and bringing them back alive in July of 1969. As a matter of fact, the Vanguard program went on to launch an advanced version of the failed Vanguard, which is still orbiting in space.

About the same time there was a great deal of interest in launching communication satellites. The first of these was COMSAT. The Comsat Corporation was created by the Communications Satellite Act of 1962 and incorporated as a public trading company in 1963. The primary purpose was to serve as a public, federally funded corporation intended to develop a commercial and international satellite communication system. With COMSAT in orbit, it was finally possible to have transatlantic phone calls and transatlantic television. There was no such thing as the internet at that time but there certainly was a transfer of data between different systems.

North Americans and Europeans were ecstatic over this capability. To give the reader some idea of this impact, let me tell you about the broadcast in North America of the coronation of

Queen Elizabeth II on June 2, 1953. The event was telecast in real-time in the UK. A Canberra bomber (UK designation – B-57 in the U.S.) carrying the exposed film took off at the conclusion of the ceremony. The film was processed in flight. The plane landed in Montréal 3½ hours after takeoff and was rushed to the broadcasting facilities of the Canadian Broadcasting Corporation in Montréal. The film was approved for broadcast throughout North America. As a result, approximately five hours after the event ended in London, it was being broadcast in North America.

With Comsat, and all the successor satellites, it was and is possible to broadcast everything in real time.

Rocketry advanced dramatically with the Mercury, the Gemini and the Apollo series and then with the space shuttle. I had some hand in this since my doctoral thesis was concerned with establishing the heat transfer characteristics as space vehicles returned to Earth from outer space. By knowing the heat transfer requirements, it was then possible to develop heat shields. The first of these was an adiabatic compound that would sublimate and in that fashion, dissipate heat. The more sophisticated approach was to use tiles that would reflect the heat and not transmit it into the vehicle. However, the danger was if any tiles were dislodged then the heat would penetrate into the vehicle and destroy it. As is well known, that happened with both the Challenger and Columbia Shuttles.

On January 28, 1986 when O-Rings separating the right solid rocket boosters failed because of the cold temperature. This caused the Challenger's rocket to rupture and explode. The entire crew was killed. The second space disaster occurred on February 1, 2003. Following a successful sixteen-day science mission, the space shuttle Columbia broke apart upon reentry, killing the entire crew. A gouge in the shuttle's left-wing caused by the impact of insulating foam from the shuttle's external tank during liftoff on January 16 caused the accident.

The tiles used on the space shuttle vehicle served their purpose, but were always a source of potential disaster. Subsequent space shuttle launches carried repair capability and crew members were trained for extra vehicular forays to repair any damage in launch.

Future space shuttle vehicles, especially those associated with carrying a large number of colonists to Mars are planned to avoid the possibility of dislodging heat reflection tiles on launch.

Satellites

From the 1960s on, numerous satellites were launched for communication purposes as well as for spy purposes. The communication satellites were of great interest. They were also very expensive.

There are three orbit systems for communication satellites. They can be set for high orbit so that very few satellites can actually cover the entire earth and they seem to be stationary above the Earth at the same point. In other words, they are high enough so that their motion keeps them in the position that they would appear to be stationary, even when they are not. This is known as a geostationary orbit (GEO). It is a circular geosynchronous orbit in the plane of the Earth's equator with a radius of approximately 42,164 km, 26,199 miles, measured from the center of the Earth. The orbit of such satellites is about 35,786 km, 22,236 miles, above mean sea level. Satellites in GEO orbit certainly have an economy of scale in the reduction of the number of satellites required to cover the entire surface of the Earth. However, they are impractical because of the delay factor (redundancy) as well as the deterioration of the signal in traveling such distances, even if they are straight-line.

The Medium Earth Orbit (MEO) is used for satellites closer to the earth. Orbital altitudes vary from 2000 to 35,076 km, 1243 to 22,236 miles, above the Earth. The region below medium orbit is referred to as Low Earth Orbit (LEO). LEO orbits range from 160 to 2000 km, 99 to 1243 miles, above the earth. Most communication satellites are placed in LEO orbit. While the number of such satellites is greater than those required in higher orbits, the redundancy is acceptable, and signal strength does not deteriorate. Signal strength falls off as the square of the distance from the source. Hence the effect of distances is significant. The cost of launch into low orbit is much less than the cost of launch at higher orbits; signal strength at high orbit must be greater than that in low orbit; and redundancy, of course, increases with altitude. Taking all factors into account, LEO orbits are the most interesting for communication satellites, especially those that will be launched in the immediate future.

The first attempt at a network was the Iridium project in 1985. This had sixty-six satellites orbiting at 700 miles. The cost of this program was in the billions of dollars and it was a total failure. By the time it could be put into use, it was supplanted. The iridium system was conceived of 77 satellites with seven orbits of 11 satellites each. Hence the name Iridium which is the 77th element in the periodic scale of elements. In actual fact, only 66 satellites were required in six orbits of 11 satellites each. The satellites weigh 519 pounds each – 689 kg. Each had two deportable solar panels and batteries. Ninety-eight satellites were built, 95 of which were launched. These launches occurred beginning on 5 May 1997 and ending on 20 June 2002. They were placed in low orbit - LEO - at an altitude of 485 miles, 781 kilometers.

Technically, the system was successful. However, because of the high cost of producing the satellites, the prices were very unattractive. Business did not develop sufficient to meet costs. The company went bankrupt, the largest bankruptcy in US history up to that time, in excess of $5 billion. The assets were acquired at a fraction of the original cost. Services by the system were re-priced but were not totally economically feasible even then.

More modern satellites were launched and cost millions of dollars or more and sometimes weighed a quarter of a ton.

In 1999, Stanford University proposed the idea of cube satellites or cubic satellites and these would weigh less than 10 pounds. They are currently being used for communication satellites and a whole bevy of these can be launched at one time and the cost of launching these has shrunk to about $3000 per satellite. It is now possible to have a personal communication satellite. Just think of what that means. Just think of the security aspects. In addition, many companies can now launch a whole series of satellites and offer services as well as controlled use for its own purposes. Amazon is one of the most sophisticated of these, as is Google. We can look to where the satellites of the future will be the CubeSats with hundreds of them in space spanning every part of the globe. As a result, at very low cost it is now possible to offer internet service and cellular phone service in all parts of the world. Deepest Africa and remote Siberia will all be fully "wired" with regard to the rest of the world.

I might tell you a humorous anecdote here. In the year 2000 I was in Russia visiting the Kurchatov Institute and was a guest of the President, Dr. Eugenio Federov. The vice president of the university happened to be an Admiral in the Russian

Navy. He showed me a handheld reactor, actually smaller than a football. He also took me on a tour of the first nuclear pile in Russia dating back to 1944. I was impressed with this research, and thought the Russians had done a tremendous job, until he told me that the whole work was financed by the US Navy. I forgot to ask him if the design was stolen from the Navy or if the Navy provided it as help for Russian economic stability following the demise of the Soviet Union in 1989-1990.

In any event, the point I am making is that satellites from the earliest beginning have now advanced to the CubeSats. Various attempts are being made to provide propulsion power to the CubeSats so that their life would be extend beyond the one and a half to two years before they would lose speed and drop into the atmosphere and burn up. These attempts are associated with magnetic induction that can create certain magnetic waves, which can keep the satellite aloft and also change, its position. Another means is to have an arc discharge utilizing the external atmosphere as fuel. The whole idea is to provide propulsion capability without any fuel having to be carried.

To date both of these examples are working and it is expected that operational models with a weight of one to two kilograms will be available shortly. Hence it will be possible to launch a carrier of litter of CubeSats, which can be ejected as the launch vehicle circles the earth or alternately they can be ejected on spot and using their propulsion systems find their pre-established orbit position.

The end result of this is to provide universal communication capability. Every part of the globe will be Wi-Fi enabled. Private Wi-Fi networks will also be possible and also personal Wi-Fi networks. Just think of what that means. At this time, there are approximately 7.25 billion people on earth. Just about a third are fully wired and connected. Now think of connecting the other two-thirds. How many jobs are created? How many teachers will be needed? How many nurses? How many of anything? Just think of the food supply needed. Will we exceed our capability of producing food? No! As a matter of fact, we can produce enough food to feed three or four times our current total world population. All we have to do is start applying modern techniques of production to agriculture. Hence for now we can say that within the next ten years every corner of the earth will be cell phone, smartphone and internet enabled.

Interplanetary Travel

At the current time, it is proposed that interplanetary travel be changed from the manner in which we proceeded to the moon. Our flight to the moon was really not a flight to the moon, but really a three-stage process. Stage number one was to escape the gravitational pull of the earth and in a sense to be flung into space. Stage number two was to coast based on that initial thrust and propulsion.

Stage three was to be captured by the gravity of the moon and then to land on the moon.

If this approach is taken for a trip to Mars, then that trip will take months to accomplish. On the other hand, if there can be continual propulsion during stage two as already outlined, then instead of coasting, the space vehicle will increase its speed. If the means of propulsion were continual, then that speed would be increased continually. Even if the acceleration component of the thrust is small, the ultimate speed of the vehicle over time will be enormous.

It is possible to utilize plasma propulsion or plasma discharge to create a propulsion capability. While this propulsion capability would be minimal, it must be remembered that it is continually additive throughout the entire trip. Hence in this fashion it is possible for a trip to Mars to be accomplished in less than a month. Depending upon the power that is available to the rocket ship, and the size of the rocket ship, conceivably that trip could be reduced to two weeks.

Plasma Propulsion

How does this plasma propulsion work? Outer space is a near vacuum. As a result, it would not be necessary to create a vacuum in order to initiate an arc discharge. A propulsion motor would then consist of an opening to the outside atmosphere, the near vacuum of space, and the initiation of an arc discharge leading to the creation of a stream of ions that would form a means of propulsion. The arc discharge would be generated by solar panels creating the power necessary. Hence the size of the space vehicle will be an important factor in the size of the solar panels. With solar panel capability continually increasing in efficiency, then the amount of power generated could be substantial. By selecting the time for the transit to Mars to occur when there is minimal night, as such, in the exposure to the sun, then it can be assumed that the trip would be quite short. Batteries would accumulate unused power to be available in darkness.

An alternate source of power could very well be a miniaturized nuclear reactor. As already indicated a little earlier, in the year 2000 I held one of these in my hand. It was the size of a football. It weighed only a few pounds – probably less than 10, and maybe as low as 5. I wasn't tuned in to consider weight at that time.

In any event, with plasma propulsion it should be possible to have interplanetary travel within reasonable periods of time. The reader may very well ask how they can be fueled in space, when space is a vacuum. But space is not a total vacuum. The density of the material filling space is only a minor fraction of the density of the atmosphere at sea level. Rest assured, there is more than sufficient gaseous molecules, even in deepest space, to provide all the fuel necessary for plasma propulsion to take mankind to the farthest points of space. To be more precise, outer space is heavily composed of nitrogen molecules. These are quite suitable for the process of arcing to create the ion stream fundamental to plasma propulsion.

Warp

There is, of course, an even more revolutionary concept and approach. That is the concept of warp, wherein time and location are considered separate. Hence, it would be possible to instantaneously change position by "warping." Currently, this is the realm of science fiction. However, even with our current limited knowledge of the exact relationship of time separated from position we can conceive of this type of position control. It is certainly an interesting concept.

Currently, the main proponents of the voyage to Mars are Boeing and Elon Musk of Space X. This latter organization has achieved success in recovery of the first stage. Currently the first stage is recovered in a barge at sea. Plans are to attempt a soft landing of the first stage back on the launch pad.

Provisioning the Space Station

To date, the space launches for the United States have been executed by ULA (United Launch Alliance), a joint venture of Boeing and Lockheed. The engine used was the Russian RD–180. This arrangement was made following the collapse of the Soviet Union in 1990. ULA is now developing its own engine, relying heavily on Blue Origin, a company owned and directed by Jeff Bezos, the driving force behind Amazon.

Space X, controlled by Elon Musk, has developed its own engine and has helped bid and capture four supply contracts into the space station. It had two disasters with delivery vehicles; but has achieved success in recovery of the first stage in a barge on the ocean.

Other engines have been developed, or are in development, by the European Space Agency, Japan, India, Russia, and Canada. The required power of the engine will depend upon its mission, and the methodology of achieving that mission. For example, for interplanetary travel, the payload, with or without passengers, can be launched directly from Earth in the usual manner, from the moon as a staging platform or base, or connected in orbit to fuel stages parked in orbit for just such a hook up.

As will be considered later in this chapter, other possibilities exist on interplanetary travel as well as for launching and maintaining small satellites – the CubeSats.

Passenger Trips into Space

Space trips are proposed for passengers to travel even as far as the moon. The major proponent of this venture is Richard Branson, the owner of Virgin Atlantic.

If current technology is used, as explained by Elon Musk of Space X, the Mars flight can take place in 110 days and hopefully in as few as 80. The approach here would be standard rocket technology. A spaceship would be launched from earth and

163

tethered into earth orbit while the launch vehicle descends to earth to be launched again with the fuel for the Mars trip. It would then be connected to the orbiting spaceship with the colonizers and proceed under power to Mars. It would be a soft landing so that the same spaceship could be launched from Mars. Conceivably other spaceships would be sent to Mars with fuel packets in orbit around Mars for the return trips to Earth. It would be necessary for the spaceship that lands to have sufficient fuel to take off and enter the Mars orbit to pick up a fuel pod for the return trip to Earth. The gravity of Mars is significantly less than that of Earth, to be more exact; it is 38% of the Earth's gravity. This would significantly reduce the amount of fuel required to escape from the initial pull of the Mars gravity in order to enter orbit around Mars to pick up a fuel pod for the trip back to Earth.

While the system of picking up fuel pods that are in orbit around Earth and in orbit around Mars will take some time to accomplish, it must be remembered that during that same period of time people will not necessarily be returning from Mars. However various ores and other items from Mars being sent back to Earth must also be taken into account. But once again, initially, this will be a small fraction of the initial weight of a ship loaded with one hundred or more colonizers.

The entire plan of Space X as outlined by Musk on September 29th of 2016 is available on the web.

Two alternative methodologies have been advanced with regard to ensuring a fuel supply for the Mars voyages. The first is to use the Moon as a staging location, and to create a fuel storage depot there. The second is to create a space station that would act as a fuel depot. Both approaches could work effectively in providing the fuel for the Mars voyages, as well as other interplanetary space flights.

On the whole, it is my personal opinion that interplanetary travel will utilize methodologies that would not require fuel to be carried. I believe variation of plasma propulsion is the answer. I felt this way since 1952 when I first saw the effect of creating a plasma stream in the shock tube. Even in those early days of the space revolution, we proposed a spaceship that would take off as a non-spacecraft, using the power of turbojet engines until it achieves an altitude of about 80,000 feet. Ramjets would then propel the passenger capsule into near space, an altitude of 50-100 miles. The plasma engine will then produce a continual acceleration component. The speed of the capsule would increase, and the trip would be a few days instead of few weeks. There is no doubt in my mind that plasma, or a variation of plasma propulsion, will be the means of achieving interplanetary travel.

On the whole the time table that Elon Musk of Space X laid out was to have a series of Mars flights starting as early as 2017. While this is quite likely to occur. It is also quite feasible, that by the time actual Mars colony flights occur, plasma propulsion will be sufficiently developed to be used. If that is so, then the idea of parking pods full of fuel in orbit around Mars and around Earth will not be necessary.

One way or the other it is certain that successful Mars flights will occur sometime in the next ten years.

This is a far cry from the initial idea that the first Mars colonizing flight would be a one-way trip lasting months. We are now down to approximately three months under normal propulsion systems, and probably a month using plasma propulsion. Conceivably that could be reduced to two weeks.

One way or another there will definitely be Mars colonization trips starting in the next ten years. On the whole, the exploration of space, and the use of space to provide complete communication capability on Earth, will revolutionize the combination of space flight and computer technology united in a common system. Moving in the next ten years will be an adventure beyond belief. Granted there will be some dislocation and disruption for the jobs of many people, but the jobs created will far exceed the jobs being eliminated, even though it won't look that way or seem that way during the first part of the transition from the old to the new.

As the inventor of the smartphone, it is my strong belief that the next ten years will be the start of unparalleled and unbelievable development in prosperity for the Earth and for all its inhabitants, even as many of them leave Earth for other planets.

CHAPTER 10
Shopping and Security

Shopping malls did not always exist. They came as an extension of the strip mall with a series of stores tied to a parking lot, which in turn came as an outgrowth of individual stores in a geographical area. In other words, it started with a single store, joined by other stores, along the street. This accumulation of stores then led to creating major parking areas, as the car became an important element in shopping. The desire to have more variety gave rise to the shopping mall. In an attempt to satisfy all ages' entertainment, dining areas, confectionery shops, and motion-picture houses were added to the shopping outlets in the mall. Along came Amazon, Walmart, Costco, and the other super electronic and bulk shops. The major difference between Amazon and all the others was in the use of the Internet. Suddenly there was conflict. Why go to the mall when you could buy everything on Amazon? That will be the major issue of the next ten years.

The mall operators have the means to compete. In the past, it came down to quality of goods, nationally known shops, and the experience of turning shopping into an adventure, a fun time with friends, and the place to go. The super mall in Edmonton even had a circus. Underground malls in Montréal were convenient to the real world commuting terminals. But still, the ability to go online and do comparative shopping without travel was certainly attractive.

This conflict will intensify.

Shopping ten years from now will be somewhat the same as it is today with some major exceptions. The ratio of ordering online versus going to shops will be inverted. In other words, in the present time more people go to the store or to the shop than use online shopping. This will be totally reversed. People will become accustomed to online shopping and travel to stores will be significantly reduced even though such travel, using either driverless car, or driverless taxis will be much more convenient than it is today. With the driverless cars, passengers disembark at their destination, and the car will park itself. Then, upon demand, the car will come to wherever you are for you to embark. This is a joy to anticipate. It is also a problem since it will open you to having your car stolen. Currently cars can easily be hacked. However, much improved security systems hopefully will prevent this.

Driverless Cars

It must be mentioned now, and it will be covered in great length in the last chapter, that hopefully it is you controlling your car. Many driverless cars will be stolen since hacking will be very profitable when it can deliver a car to you. Hacking is and will continue to be a universal threat. Countermeasures will be described later in this chapter. The seriousness of this problem cannot be understated. It is not only cars that can react to the hacking threat, but also Heart Pacers, insulin machines, and any kind of monitoring system. If there is a control chip or board linked to a wireless capability or network, then it is open to hacking. On the other hand, hacking can be prevented and stopped in its tracks.

For now, assume that there are significant protections that make it impossible for your car or health monitoring device to be hacked. No matter where your car parks itself, you will be dropped at the entrance to a parking mall, and picked up at entrance, and not necessarily the one you were dropped off at, How convenient!

Just think of what that could mean to the mall operators. Very simply put, the car parks itself. Hence, where it parks is not an issue. When a human parks the car, the desire is to be as close as possible where you want to be. This gave rise to huge parking lots surrounding shopping malls. Remove that requirement, and two things happen. First of all, parking need not be in close proximity to the stores; and second, the parking lot in close proximity to stores can be used for other purposes. For example, a circus, launch aprons on trips to space, or special outdoor exhibits.

Most certainly the parking lots of malls will be used for these purposes as more and more cars are equipped to be driverless.

In any event, proceeding to a mall will be quite convenient, but the incentive to shop there will be significantly reduced. Let me explain further.

Using augmented or virtual reality, it will be possible to investigate every store in any mall whether the store exists or not. Furthermore, the products can be handled, and through sensory gloves that can be donned, even cloth can be felt in terms of the type of reaction you would have by fingering and rubbing together the actual cloth.

Through the dual frontal camera arrangement or the newer Smartphones, it would be possible to see yourself in virtual reality, wearing any garment or sweater that you wish to try on. The trial would be calculated according to size of the garment and knowledge of your body dimension. An AI program will "fit" a sweater or even a suit to an image of your body, and show you the appearance. Any alterations could also be calculated and automatically marked.

Hence, all the functions that you would perform by going to the mall can now be done without you going to the mall. Shopping would be much more convenient by doing it electronically.

The Business of Malls

It would also be cheaper. There is no doubt that the cost of merchandising would be significantly lower if the cost of a store were eliminated. Not only is the rent high, but the personnel cost must also be significant. All of this would be eliminated when the shopping is done

171

electronically. As a result, electronic malls would be very rapidly instituted, and to some extent these will replace the actual malls.

There will be some attenuation or reduction of mall expansions. Malls will not disappear because they are meeting places as well as shopping places. This is especially so for the younger people. The malls will become more of a meeting and entertainment location. Sports events and various entertainment functions will begin to be scheduled for the malls. The complex of each mall will become a mecca for the younger shopper. The mecca will also exist for the older shopper, but with a different motif. I suspect square dancing, old-time music concerts, and sports events will be attractive to the middle-aged and aged.

The counterattack by malls will be to provide something of interest for all age levels, turning the mall into a place to be and not just a place to buy. In fact, the malls will probably be able to work arrangements with government agencies, having them underwrite many of the entertainment activities in the malls. This would certainly be less expensive to the government agencies then constructing new stadiums or new entertainment centers.

Sports attractions, exhibitions, circuses, merry go round type of rides, and sports encounters of every imaginable type will be scheduled for the super malls of the future. Since travel to the malls will be relatively simple with driverless cars which will be able to drop you at the entrance to the mall of to a specific sporting exhibit, and pick you up no matter where you are, thus simplifying the travel, and making it more attractive for you to go to the mall to attend a function that you want to be at or see.

Hence, malls will become more exhibit locations with entertainment for the masses; and shopping will be switched to electronic malls and such volume in physical malls will be much reduced. You can conceive of the day where there will be no physical shopping in the malls what so ever. It'll probably be somewhere close to the ten-year period.

There's one great advantage to having electronic shopping. That is that the item you select can be delivered to you the same day. The delivery will be by drone, or by robot. As this book is being written, the use of drones for delivery is being pioneered in the UK by Amazon with special arrangement by the government of the UK. It is expected that this test will be successful. The idea I would advance is that drones will be permitted to fly between 50ft and 100ft above the ground, but they must follow existing roads in order to move to different locations. That way there will be some control over drone traffic, ensuring it would not interfere with airplanes, even those that are landing. By regulating drone traffic to follow the same traffic patterns as existing highways and roads, while the travel time might be longer than if the drone flew cross country as such, it would be regulated and the drones would not collide with each other, or with aircraft. Furthermore, the drones would be regulated by insisting they follow the same stop and go patterns as vehicle traffic at ground level.

And so, shopping in the future will be materially simplified. This would apply to clothes shopping as well as grocery. Groceries would be delivered to your home the same day you have done the shopping. Oh, what a joy this would be when you run out of cream for your coffee at midnight. You would dial up your Smartphone, and pick the cream that you want and it would be delivered to your home by drone or robot within an hour.

With this convenience for shopping, it is expected that shopping would increase even though the method of shopping would be radically different. With this simplicity of purchase, people would certainly buy more. With this ease of returning as well, more goods will be returned. But just as the inner lines have moved to a nonrefundable ticket, so too the merchants will move to a non-returnable purchase arrangement. Doing that will stabilize the purchasing agreement so people will buy only what they really want to keep, and when they do so, it will be with the provision with not being able to return it, so they will spend much more time in perfecting their selection, which will take much less time since it will be done electronically. On the whole, shopping will be easier, be more productive for the buyer, and more profitable for the seller. Similar to the extensive and rapid development of Amazon, so too the shopping malls will expand rapidly with the merchants who can guarantee a wide range of products through their arrangements with many manufacturers as well as those who can guarantee rapid deliveries.

Shopping in the next ten years will certainly be different than it is now.

Security

In my opinion, much of the hacking recorded in the press these days is not hacking at all. I think a great deal of it is hardware related. It would be a very simple matter for a manufacturer of chips or computers to insert a "backdoor." On command from this backdoor you could download data from the system, or permit the operation of a program. This backdoor approach can be neutralized in many ways. The system could be flushed prior to being used to eliminate any dormant programs or back doors. Alternately, the operation of the valid programs of the system can be hardened so that only such valid programs will be followed or executed by the system. Other alternatives are for the valid system programs to be encrypted, contain special verification coding, or be made impervious to additions. There are many ways in which the back-door attack can be thwarted.

One other source of external hardware hacking that can occur is via thumb drives. Beware of special sales of thumb drives. It has been known that thumb drives can contain dormant programs that can be loaded into an existing system for mischief, or can download data from the system. It is strongly urged that any new drive be flushed to ensure that using it will not create a backdoor for external control of your system.

The Stuxnet Situation is one of the most famous cases of thumb drives being used to penetrate a high security system. The system in question is the Iranian Nuclear Development Program wherein a large number of centrifuges were being used to produce enriched Uranium for a bomb. A thumb drive purportedly was introduced that changed the spindle speed of the centrifuges and set the whole program back by months. Beware!

A favorite trick of the hackers is to hoodwink someone into revealing their login ID or their password. With knowledge of the email address and the login ID, the hacker can secure a new password that will provide access to the system, and even lock out the legitimate user. To guard against this possibility ISPs customarily send notices to users concerning any change to their login credentials – login, password, notification addresses or phone numbers. By the time such notices are received it is often too late.

Another favorite trick of the hackers is to pose as a bank or other legitimate organization and dupe the user into revealing information. Hence, in order to thwart hackers and guard the security of your system it is strongly recommended that:

1. The email address you use for login purposes is never used for anything else; in other words, maintain a series of email addresses one or more of which should be treated as "Top Secret".

2. Never reveal your login name.

3. Change your password and login name periodically, with the periodicity shorter for highly sensitive systems. In all cases, make your password and login ID complex. Never use personal data and names to create your password or login ID.

4. Never write your login information on a Post-it note and attach it to the face of your computer.

5. Never let anyone use your system after login.

6. Follow strictly all the administrative guidelines for security.

During the next ten years, more and more electronic capability will exist, offering even greater opportunity for hackers and thieves. However, the problem is being addressed and solved by adding sign-ins connected to physical attributes of the user such as fingerprint, voice, face imprint, DNA, etc. Hence no device that can affect your body will have any possibility of being penetrated. At the very least it will give you control as to who has control over your system.

With regard to the type of hacking that is reported every day at this time, hopefully that too will be over. Many developers, including myself, are developing a methodology currently being incorporating in software that will make it impossible for a hacker to be successful. Hackers, then, will rely on gross negligence on the system administration. Let me repeat. It will be impossible to hack unless a target of the hacking is guilty of gross negligence and does not pursue the simple protection measures required.

SWIFT (Society for Worldwide Interbank Financial Telecommunication), the global financial messaging system, disclosed new hacking attacks on its member banks as it pressured them to comply with security procedures instituted after the February 2016, $81 million theft at Bangladesh Bank. By a fluke, the effort of the thieves to steal about $950 million was thwarted. After reports of previous incidents SWIFT prompted regulators in Europe and the United States to urge banks to bolster cyber-security.

The Bank of England recently ordered UK firms to detail actions to secure computers connected to the SWIFT system, while the European Banking Authority said domestic authorities should stress test banks for cyber risks. The Federal Reserve and other U.S. agencies told banks to review protection against fraudulent money transfers.

In a recent private letter to clients, SWIFT said that new cyber-theft attempts have surfaced, when it last updated customers on a string of attacks discovered after the attack on the Bangladesh central bank. The disclosure suggests that cyber thieves may have ramped up their efforts following the Bangladesh Bank theft, and that they specifically targeted banks with lax security procedures for SWIFT-enabled transfers.

The Brussels-based firm, a member-owned cooperative, further indicated that some victims in the new attacks lost money, but did not say how much was taken or how many of the attempted hacks succeeded. It did not identify specific victims, but said the banks varied in size and geography and used different methods for accessing SWIFT. All the victims shared one thing in common: a weakness in local security that attackers exploited to compromise local networks and send fraudulent money transfers.

Accounts of the attack on Bangladesh Bank suggest that weak security procedures made it easier to hack into computers to send SWIFT messages requesting large money transfers. The bank lacked a firewall and used second-hand, $10 electronic switches to network those computers, according to the Bangladesh police.

SWIFT has repeatedly pushed banks to implement new security measures rolled out after the Bangladesh robbery, including stronger systems for authenticating users and updates to its software for sending and receiving messages. But it has been difficult for SWIFT to force banks to comply because the nonprofit cooperative lacks regulatory authority over its members. SWIFT told banks that it might report them to regulators and banking partners if they failed to meet a deadline for installing the latest version of its software, which includes new security features designed to thwart the type of attacks described in its letter. The security features include technology for verifying people's credentials; stronger rules for password management; and better tools for identifying attempts to hack the software.

It is recommended that anyone who doubts the seriousness of Cyber Crime should do an exhaustive search of such crimes on the web. The results will be sobering. Large sums of money have been stolen. Unless controls are instituted, and followed, there will be more.

I well remember my shock back in the 1970s when I was developing systems for electronic transfers for vast sums of money. I built significant protections into these, and to my absolute startled surprise found that in many instances these protections were not being used because they took time and quite often the operator could not be bothered. When I insisted that they be used they were rigidly enforced with my clients. I can safely say that none of my clients were ever hacked. I can also relay that we worked with our clients and quite often hired hackers, who didn't know they were working for us, and asked them to hack into our systems, giving them full access. We then watched very carefully everything they did and just plugged the holes behind them. We made it impossible for any hacker to get into our systems. This became rather important in the latter part of the 20th century, the volume of financial traffic through our systems was close to three trillion dollars per day. Some of the transactions exceeded 1 billion dollars. These were prime candidates for hacking. No one got in!

It is possible to develop systems that are hack proof. We did it. We're doing it again.

Security Support

An affective security system can only exist if administrative controls are rigidly adhered to. It is the nonuse of such controls that would make penetration of the most secure system possible.

For example, it is standard practice to ensure that improper use of highly secure equipment by operators who are not cleared for the use of such equipment is made impossible. Otherwise it would become an open invitation to fraud, and significant loss, if not destruction of the system.

For example, in the creation of many funds transfer systems; the operators of these systems capable of transmitting control of vast sums of money are normally separated into cages. Access to these cages, and resulting access to the computers, is highly restricted only to those cleared for such activity.

Another administrative situation is to have three people involved in the transfer of money. One to enter the original data. A second person to reenter the data without knowledge of what has been entered by the first person. This second person would also have the ability to edit or change. And by the way any time any change is made to any record, an audit trail is made to that in terms of the new value, the old value, and who did it and why and when.

The third person would have the ability to release existing wire transfers, but would not have the ability to enter or change. Such a person would, of course, have the ability to cancel an existing wire. In such an event, a full audit trail is made of the cancelled wire, including the person who cancelled it, the date and time.

Currently, most wire transfer systems, give third party, or users, the ability to enter all the details associated with a wire transfer. These can be "standard" wires, where only the amount has to be entered for it to become effective. Having created many of these systems, on a first-hand basis I can say this is very dangerous. Unless there are adequate controls in terms of how these are set up and monitored, creating this ability should be discouraged. If it is allowed, then there should be adequate safe guards to ensure that the party sending the wire has the full responsibility for any malfunction or theft associated with this process.

In the future, there will be many forms of hardware, software, and other access control procedures. All of these might appear to be restrictive, but should be instituted, and controlled in terms of performance if security is to be maintained. Remember that the amount of money transferred in any single day is often multiples of the total asset value of the bank. With regard to third party transfers, it should also be remembered that quite often these transfers are also a significant percentage, if not multiples, of the total asset value of that company. As a matter fact, more importantly, even if there are adequate controls, it should be remembered that every attempt would be made by thieves to circumvent all of these controls.

Note that there is no limit to the amount that can be diverted or stolen. There's further no way that the trail can lead to a successful recoup. Quite often a trail will end as a dead end. For example, many of the casinos in the Philippines of Macau cannot be penetrated in terms of the disposition of funds that enter their accounts. The same is true of many situations in different locations around the world. Casinos, of course, are notorious. But so are many of the financial havens, most especially in some of the smaller countries of the world.

I would hazard a guess, if not a wager, that millions are stolen each day. Beware!

CHAPTER 11
Surviving Technological Changes

Life in a connected world will be different. It will certainly be more convenient since robots and various devices will do more for us. Cars will be driverless and will pick us up where we are and deliver us to wherever we wish to be. Trips will be enjoyable as we will be able to work if we are alone, or talk in a living room center if we are with others.

Life will also be easier from the point of view of less work to do no matter what your profession. There will be more public entertainment, as the government will strive to occupy the time of its good citizens so that they will not engage in mischief or revolutionary activities. Police activity will be much more stringent, and there will be significant incentives to become part of the government either in the law enforcement or in the military.

The military to a large extent will be a branch of law enforcement, on a national as well as an international scale. Warfare as we know it today will be significantly reduced. It will become increasingly electronic. Much of what everyone wants will be provided, much of it by mechanisms that will be available to all at a relatively low cost. Many of the day-to-day costs of life and government will be significantly reduced. Hence, the cost of providing entertainment and increased law enforcement will be offset by the significant reductions in the cost of government. The cost of health care will certainly be a reduced burden on any government, since it will be provided at a much lower per patient cost than currently because of the vast deployment of robots and of electronic monitoring, diagnosis, and treatment.

Major economic disruptions, however, will occur because of the changing nature of many of the major industries in the world. Foremost of these will be the petroleum industry. It will be materially affected, and dramatically so, with the electric car. While gradual, this impact will build until the demand for oil or fuel becomes only a fraction of what it is today. The price of oil per barrel will plummet. This will obviously impact the economies and power of many oil-producing nations. Some will not be seriously affected because they have other forms of national income. Some, however, such as Saudi, Iraq, Iran, and the Middle East generally, will be seriously affected. These losses of income in oil production can be offset by income from educational programs, airlines, resorts (in places like Dubai) and financial investments. There is no need for any nation that uses its financial resources to innovate to suffer as the scope of world use of oil changes.

Implementation

The rise of electric vehicles and alternative fuels such as solar power, the reduction in the use of petroleum products for transportation and for generated municipal power will seriously impact the use of oil. Energypost.eu shows the anticipated rise of the alternative energy vehicle with the concurrent decline of the internal combustion vehicle.

The current oil production on a daily basis is approximately 48 million barrels. Of this, approximately two thirds is produced by a small number of nations- Iran, Saudi Arabia, Iraq, Russia, Canada, and the United States. The cost of production for one barrel of crude oil ranges from a low of $8.50 in Kuwait to a high of $52.50 in the United Kingdom.

With the exception of Canada and the United States, it is expected that these nations could consider efforts to delay progress towards eliminating oil as a fuel for cars.

In the next ten years, more new cars will become increasingly electric. This will probably be most pronounced in India and China, although a significant number of electric cars will be produced by Tesla Motor Corporation in the United States. The Tesla car has been successful but with the introduction of the zinc ion battery this should be accelerated. In addition, the introduction of the SUV version and the low-cost version of the Tesla car will increase the number of such cars introduced. Projections are that the volume in the next three years will rise to more than half a million electric cars per year. At the current time, every Tesla car produced is equipped with driverless capability. The projections for this company are aggressive.

Other car companies are engaged in producing electric cars. This is especially true in India and China. Small electric cars, with or without driverless capability are expected to dominate the car market in those countries, and in most of Asia. The expected increased demand for petroleum products in those countries will not occur.

In the Western world, the production of the electric car will be coincidental with the driverless car, as mentioned earlier. General Motors and Ford are aiming to have a car with no steering wheel or brakes. Volvo has a fully driverless vehicle that is in active testing at this time in Pittsburgh with the Uber Corporation. As already indicated, all Tesla cars produced now have driverless capability.

The impact of electric cars on oil production will significant. At this time 62% of the petroleum products generated are used for transportation. It can be expected that this will be reduced by a bell curve over the next ten years until total oil production for surface transportation will be reduced to less than 38% of what it is now. In other words, 62% or 29 million barrels per day might no longer be necessary ten years from now.

The economic factor of this will be significant. There are many situations where oil production does not pay if the cost of oil sold is less than $50 per barrel. The UK for example requires $54 per barrel to produce profitable oil. Saudi Arabia requires only $20 per barrel. Many wells in the United States only require less than $20 a barrel, but some of the newer refineries, especially those associated with shale or tar sands, need $50 a barrel. In the case of the United States this is no problem since government subsidies can be justified on the basis of national security so that in the event that oil falls to $20 per barrel, US production could continue.

When the same principles are applied to the use of petroleum productions to generate power for municipal and industrial use, it becomes even more dramatic in the means of reduction. Currently 24% percent of oil produced is used for the generation of power for industrial and domestic use. If solar power panels are used for all the homes in the United States and other areas, and if solar power and other forms of recurring power such as wind and hydro power are used to the fullest extent, then it can be expected that 12 million barrels of oil per day will no longer be needed. This will have an additional traumatic impact on the production of oil. This in turn would significantly reduce the price. It is expected that the total production of oil might very well fall from 48 million barrels per day currently to less than 10 million barrels per day in the next ten years. It is further expected that the price will probably fall to less than $20 per barrel. This will have a completely catastrophic effect upon most of the OPEC nations, destroying the economy and wiping out their reserves. It is expected that this will have no impact upon the United States, since, it is hoped that the government will be far sighted enough to institute support for petroleum generating organizations so that the United States will continue to be petroleum independent from the rest of the world.

Electric cars are finally nearing the type of performance desired. This consists of a range of between 400 and 600 miles between charges. This has been achieved with a Tesla automobile, which is a true electric car. This automobile has relied on the lithium-ion battery to date. However, advances currently acceptable with zinc-ion batteries will make this a suitable replacement for the lithium-ion battery. This will eliminate the danger of unnecessary heating and resultant fire. At the time of writing of this book, the lithium-ion battery in the Samsung Note 7 cell phone is a disaster. Apparently, the fires are caused by excessive heat generation when the cells on the batteries are too close together. Without proper spacing and ventilation, this excess heat problem could not be resolved. The battery compartment will undoubtedly be increased in size in the redesigned Samsung Note 7. In any event, it appears that the battery problem has been resolved with the switch to the zinc-ion battery from the lithium-ion battery. The Tesla cars in the future will be made with the zinc-ion battery and the range is expected to be 400 to 500 miles.

The zinc-ion battery is also being used by electric utilities for the storage of power to provide continuous power when there is no solar exposure, such as at night. On the whole it would appear that electric powered cars finally arrived at suitable performance characteristics to supplant the petroleum driven car. Similarly solar power, with back up battery capability of significant capacity for electric utilities has now made solar power a suitable means of generating industrial and residential power requirements.

It is assumed that there will be a gradual transition from the petroleum powered car to the electric powered car. Assume that this will take ten years before 50% of all the cars are electric power. In addition, assume that it will take ten years before 25% of industrial and residential power is generated by Solar panels and not by petroleum products.

As this book went to press, India announced construction of the world's largest solar power plant that covers 10 square kilometers. When fully operational it will have capacity for 150,000 homes (648 Megawatts). At a cost of $679million, the plant was built in eight months. It is maintained and cleaned by solar-powered robots. This plant is part of India's program to power 60 million homes by solar power as part of its program to produce 40% of its power from non-fossil fuels by 2030. It is certain there are other similar projects underway or planned in other countries.

This is only a straw in the wind. Reductions in oil use from electric cars and electric power will have a major impact on oil production.

Currently it is projected that 48 million barrels of oil per day will be produced. Just think of the impact of reducing that by 25% or 50% in total. The price of oil will certainly drop dramatically considering these factors. The cost of producing oil in some older oil companies is over $60 or $70 per barrel. The cost of producing oil in the United States is $40 to $50 per barrel. Oil in Saudi Arabia has more modern oil fields as between $30 and $40 per barrel. According to the Center for Climate and Energy Solutions global transportation accounts for 62.3% of petroleum consumption. (http://www.c2es.org/energy/source/oil)

According to Bloomberg, electric vehicles could displace the equivalent of 2 million barrels of oil a day by 2023. This seems low. (http://www.bloomberg.com/features/2016-ev-oil-crisis/).

The NASDAQ website states, "for oil markets, the conclusions should be alarming." Bloomberg New Energy Finance (BNEF) sums it up succinctly: "The electric vehicle revolution could turn out to be more dramatic than governments and oil companies have yet realized."

But BNEF sees EVs displacing 2 million barrels per day (mb/d) of oil demand as early as 2023. That is just the start. The real pain will come after that point as EV sales start to skyrocket. BNEF estimates that EVs could capture 35 percent of the market by 2040, which would displace 13 mb/d. For an oil market, currently in tatters because supply is exceeding demand by a meager 1 to 2 mb/d, the destruction of 13 mb/d of demand should be unsettling, to say the least. EVs present an existential threat." Read more at: (http://www.nasdaq.com/article/electric-vehicles-could-soon-reduce-oil-demand-by-13-million-barrels-per-day-cm585937#ixzz4NYgIvKOL)

Unless there are government subsidies, it would appear that most of the oil that can be produced at a cost of less than $30 a barrel will come from Saudi Arabia then, as well as the United States. Now consider the economic disruption to the countries who cannot produce oil for $30 a barrel or less. What will they do with their oil, and how they handle their absolute bankruptcy? This could lead to international tensions or to a concerted program of such nations to find other means of generating income. Dubai has shown the way for such a path.

There is no way that the move toward electric cars and solar panel generation for residential and industrial power will stop. This will just accelerate. This will place increased financial pressure upon nations to carry the cost. Ultimately the need for oil for industrial electric power generation and for motorcars will be reduced substantially, perhaps by 50% or more. Then what happens?

There's little point in producing charts and graphs for such potential change since they will depend heavily upon the assumptions made. The assumptions that would be made are evident from the descriptions of the various paragraphs chapters of this book. The fact of the matter is that to a large extent the cost will depend on the speed with which new developments are deployed, with existing connected people but most especially those still to be connected through the deployment of satellites that will lead to the entire world being connected.

Impact of the Future

The impact of adding four billion more people to the healthcare systems of the day, and finding jobs for all those who will be displaced will certainly be major programs of the next ten years. Perhaps I am a cock-eyed optimist as the saying goes, but I believe that the number of jobs produced will be greater than the jobs that will be eliminated. I firmly believe that millions of jobs will be eliminated, but I also firmly believe that millions of jobs will be created. I also firmly believe that these jobs will not be as onerous as much of the work level is today. I firmly believe the cost of living will dramatically drop, and be well within the earning capacity even with reduced work levels. I think that most governments will begin offering a Guaranteed Minimum Wage that will be distributed by checks much the way Social Security is distributed today.

In the United States, there was some concern that Social Security will run out of funds somewhere in the next twenty or thirty years. I do not believe that will happen. I further do not believe that the healthcare plans will run out of funds. I think the cost of providing significantly better healthcare will be lower with the system as outlined in the chapter on healthcare.

I think that farming will also become much more productive and the cost of farming will be much lower than today. On the whole, I think that food production will be significantly higher, well within the requirements for the entire population of the entire connected world.

And so, we look for optimism for the future. I have been involved in the computer revolution, as such, for almost 70 years now. I have had a significant hand in many of the developments that have occurred to date, the most significant of which is the invention of the Smartphone.

I have enjoyed the adventures of those years. I personally have never had a dull moment. I don't expect to have any dull moments in the next ten years, or to be honest in the next twenty years. I look forward to these with tremendous anticipation. I hope you do too. I'm sorry that you may have lost your job. May that job rest in peace. But I am sure that you will have much more satisfaction in your life than when you had that job. So, let that job rest in peace. Meet the challenge of tomorrow. It will be exciting and rewarding.

Go for it!

ABOUT THE AUTHOR

Dr. Rocco Leonard Martino is a world traveler as an international financial consultant, computer specialist, and space pioneer. He is the published author of five previous novels, and twenty-six nonfiction books. The inventor of the 'Smartphone', he has been knighted by the Vatican, and has served on the boards of numerous international organizations, including the Foreign Policy Research Institute, the World Affairs Council, the Gregorian University Foundation, and the Vatican Observatory Foundation. By training he is a rocket scientist.

In business and scientific pursuits, Dr. Martino has been in the forefront of computer applications and process innovation almost from the inception of the computer. His contributions have created products and jobs, promoted economic growth, and enhanced the ability of technology to address important aspects of business, commerce, and government; and to improve the quality of life. Of particular interest is Dr. Martino's invention and patenting of what has become the Smartphone. Prior to organizing his own company, Dr. Martino directed the Aerospace Division of Adalia, Ltd, a firm headed by Sir

Robert Watson Watt, the inventor of Radar; directed all activities in Canada for UNIVAC, and worked with Admiral Grace Hopper on automatic programming systems; formed a partnership to create Mauchly Associates with John Mauchly, the co-inventor of computers, and spearheaded the Critical Path Method created by this company; and finally headed the Special Projects Group of Booz Allen and Hamilton. As Founder, Chair and CEO of XRT, Inc, a company he founded in the 60's, Dr. Martino served some of the largest companies in the world. Before selling the company in 1997, this company was moving some $3 trillion dollars per day through systems he designed with thousands of clients in 51 countries. These included the largest companies in the world, such as most of the oil companies in the world, the Bank of Tokyo, Prudential Insurance, ManuLife, the FDIC, US Postal Service, and the US Navy. From 1997 onwards, Dr. Martino concentrated on his invention of the Smartphone; writing books; and most recently in creating secure systems for complete protection against hackers, identity theft, and asset theft.

His accomplishments during some 65 years in technology have given him a global reputation.